STARVED FOR SCIENCE

Harvard University Press

Cambridge, Massachusetts, and London, England

Robert Paarlberg

STARVED FOR SCIENCE

HOW BIOTECHNOLOGY IS BEING KEPT OUT OF AFRICA

Printed in the United States of America

First Harvard University Press paperback edition, 2009.

Design: Marianne Perlak

pp. ii, iii: Northern Mozambique. Brian Atkinson/Acclaim Images.
p.vi: Girls carry basins of water at village in southern Niger.
Reuters/Finbarr O'Reilly.

Library of Congress Cataloging-in-Publication Data

Paarlberg, Robert L.
Starved for science : how biotechnology is being kept out of Africa / Robert Paarlberg; foreword by Norman Borlaug and Jimmy Carter.
p. cm.
Includes bibliographical references and index.
ISBN-13: 978-0-674-02973-6 (cloth : alk. paper)
ISBN-13: 978-0-674-03347-4 (pbk.)
1. Agricultural biotechnology—Africa. 2. Crops—Genetic engineering—Africa.
3. Agriculture and state—Africa. I. Title.
S494.5.B563P33 2008
630.96—dc22 2007045745

CONTENTS

FOREWORD

Norman E. Borlaug

Jimmy Carter

Robert Paarlberg is a leading scholar on the role of science and technology in smallholder agricultural development in Africa, and in particular, how governments and civil society have responded to such innovations. In this book he shows how a recent withdrawal of donor support for modern agricultural science in Africa, plus outright opposition to new farm science on the part of some global pressure groups, is contributing directly to the continued growth of poverty and hunger in rural Africa.

We share with him his concern about the lack of support for science and technology in Africa. According to the 2008 World Development Report, which this year features agriculture, Africa has only increased its agricultural research investment by 20 percent over the past twenty years, compared to a threefold increase in Asia. Whereas science has been used to boost farm productivity in Asia, raising rural income at the same time, smallholder farmers in Africa have remained starved for science. Less than one-third of Africa's cropland is planted to improved seed varieties, compared to 82 percent of cropland in Asia, and average cereal yields in Africa are less than one-third as high as in Asia.

In the early 1990s, Paarlberg was among the first to sound the alarm about the deadlock between agriculturalists and environmentalists over what constitutes "sustainable agriculture" in the Third World. This debate has confused—if not paralyzed—many in the international donor community who, afraid of antagonizing powerful environmental lobbying groups, have turned away from supporting science-based agricultural modernization projects still needed in much of smallholder Asia, sub-Saharan Africa, and Latin America. This confusion and paralysis is now most obvious in the case of agricultural biotechnology. The science of genetic engineering has significant potential to help rural Africa, particularly since it can now speed the development of crop varieties better able to tolerate stress factors such as drought. Yet the governments and nongovernmental advocacy groups of most prosperous countries, particularly in Europe, are resisting the introduction of modern agricultural biotechnology into Africa.

In 2000 a joint U.S.-European Union Biotechnology Consultative Forum was appointed by the presidents of the United States and the European Union to look at the full range of issues that have polarized thinking about biotechnology, especially in food and agriculture, on each side of the Atlantic. Although significant differences of opinion existed—mainly related to the regulatory structure involved with certifying agri-biotech products—most of the twenty U.S. and European experts on the panel agreed that agricultural biotechnology holds great promise for dramatic and useful advances in the twenty-first century. In effect, it confirmed the views of the most prestigious national academies of science in North America and Europe (including the Vatican), none of which found any new risks to human health or the environment from any of the applications of crop biotechnology commercialized so far, and all of whom confirm the potential of genetic engineering to improve the quantity, quality, and availability of food supplies.

Even so, the debate about the safety and utility of genetically modified (GM) crops continues to grow, enough to discourage governments in Africa from approving the technology for commercial use. This is a rich-world argument that is hurting the poor. Although there have always been those in society who resist change, the intensity of the attacks against GM crops from some quarters is unprecedented, and in certain cases, even surprising, given the potential environmental benefits that such technology can bring by reducing the use of pesticides.

In the past five years, the presidents of several African countries facing widespread drought, crop failure, and hunger have even banned the distribution of donated maize from the United States as food aid, having been told by antibiotechnology groups that this food was "poison" because it contained genetically modified kernels. Based on such misinformation, they have been willing to risk thousands of additional starvation deaths rather than distribute the same maize that regulators approve in the United States and that Americans have been eating for more than a decade with no documented ill effect.

Paarlberg shows that low-income, food-deficit nations are being advised by governments and pressure groups in privileged nations to reject agricultural biotechnology mostly because this is a technology the rich countries themselves do not happen to need. When it comes to new applications of medical science, which prosperous countries still need and value, genetic engineering is not seen as a threat, and is not regulated to death. Only in the area of agriculture, where new science is no longer needed by the rich—because their citizens are well fed and their farmers already highly productive—are stifling regulations imposed.

These inconsistent views regarding the use of transgenic crop technology in Europe and elsewhere might have been avoided had more people received a better education in biological science. This educational gap, which has resulted in a growing and worrisome ignorance about challenges and complexities of agricultural systems, needs to be addressed without delay.

Privileged societies have the luxury of adopting a very low-risk position on the GM crops issue, even if this action later turns out to be unnecessary. But the vast majority of humankind does not have such a luxury, and certainly not the hungry victims of wars, natural disasters, and economic crises.

The policy debate about the suitability of biotech agricultural products should focus less on risk—since after more than a decade of commercial experience with the technology, no new risks have yet been documented—and more on access for the poor. Access to biotech seeds by poor farmers is a dilemma that will require interventions by governments and the private sector. Seed companies can help improve access by offering preferential pricing for small quantities of biotech seeds to smallholder farmers. Beyond that, public-private partnerships are needed to share research and development costs for "pro-poor" biotechnology. Of course, there is nothing magic in an improved variety alone. Unless that variety is nourished with fertilizers—chemical or organic, ideally both in combination—and grown with good crop management, it will not achieve much of its genetic yield potential.

African governments, following the lead of Europe, have so far resisted the use of modern crop biotechnology. Africa has already missed the industrial revolution and the tractor and fertilizer revolution. As things stand today, Paarlberg shows, there is a risk it will miss the biotechnology revolution as well. This would be tragic, since Africa, with the largest proportion of its population engaged in agriculture, has the most to gain from biotechnologies that protect crops from disease and insects, increase yield stability under drought, enhance nutritional quality, and lower production costs.

Responsible biotechnology is not our enemy; hunger and starvation are. Without adequate food supplies at affordable prices, we cannot expect world health, prosperity, and peace.

Norman E. Borlaug, agricultural scientist, and Jimmy Carter, former U.S. president, are both Nobel Peace Prize winners who have worked together over the past twenty-two years to bring a Green Revolution to African agriculture.

PREFACE

I have enjoyed visiting farms since I was a young boy growing up in Indiana in the 1950s, living in a small university town surrounded by cornfields. Not too many miles away was the 160-acre farm where my own father had been raised during the hard times of the Great Depression in the 1930s, before he left to attend college and then become a university professor. On many holidays my older brother and I would be driven back to the farm, where we got to run in the fields, ride on the big diesel tractor, and build forts out of bales of hay in the barn. We threw green apples at each other back in the old orchard, kept a respectful distance from the livestock, and marveled at the multiple skills of our farm-dwelling cousins, uncles, and aunts. After a big holiday dinner the men from our extended Dutch family would retire to the back porch, light their cigars, and begin talking—bragging, actually—about the price they just got for their hogs, how well the corn was doing, and the new pickup truck they were planning to buy. They also offered ritual complaints about the weather, the lazy neighbors, and of course the government. Yet money was coming in, and nothing could hide their satisfaction. They suspected their sons might find a job in town or go off to college instead of staying home to farm, but they knew the value of having such a choice and were pleased they had been able, through hard work, to provide it.

When I visit farms in Africa in the course of my current research and consulting work, quite a different picture emerges. Farm visits in Africa take some planning, since a serious trip into rural Uganda, Benin, Cameroon, or Kenya usually calls for a sturdy vehicle with a full tank of gas and a reliable driver who knows both the roads and the local languages. Heading upcountry in Africa is always exhilarating, given the unmatched dramatic beauty of the big skies, red soils, and arid vistas, but soon the roadside power lines have terminated and the two-lane tarmac road has narrowed to just rutted gravel and dirt, badly potholed during the last rainy season. The driver finally stops. You walk a short distance through the dry bush until you come upon a tidy cluster of mud and stick dwellings surrounded by a patchwork of small- to medium-sized farm fields. A youngster tending goats signals to his mother that they have visitors.

Those who meet you and invite you to a seat in the shade are curious and eager for a visit, like farmers everywhere. Most are women, since the working-age men have temporarily gone to look for better earning opportunities away from the farm. It is immediately clear that little money is coming in. The school-aged children are badly clothed, and obviously not in school. The small crop of maize plants you see are struggling on cracked soil, and the cassava is withered, apparently from a virus disease. The cow kept for milk is stunted and has untreated sores. These farm women confirm they do not plant hybrid seeds or use chemical fertilizers. They have no pumped-in water. Nor any electrical power. Nor any machinery. The nearest all-weather road is several kilometers away, so nearly all that they bring in and take out must be carried by foot.

These smallholder farmers in Africa are poor—like their parents and grandparents before them—because the productivity of their labor in farming has not yet been enhanced by any of the modern applications of science that farmers elsewhere have used to prosper. No improved seeds from crop science. No nitrogen fertilizers. No animal vaccines. No electric pumps. No improved transport systems. These African farmers are laboring in an environment just as unimproved by modern science—and hence just as poor—as most farmers in Asia two or three generations ago, or for that matter most farmers in Europe six or eight generations ago. It was only when farmers in Europe and Asia gained access to science-improved farming technologies that their productivity and hence their incomes began to rise. This science-based escape from rural poverty has not yet taken place in most of Africa.

The essential role agricultural science must play in reducing rural poverty is poorly understood today, especially in rich countries, and as a consequence the idea of using technology to boost farm productivity has fallen widely out of favor. I decided to write this book after too many years of hearing comments from friends and students who were hostile toward applications of science in farming. They had somehow come to see science in farming as bad for people, communities, and the environment. Science in agriculture was feared as a harbinger of unhealthy foods, a destroyer of family farms, a precursor of market domination by corporate agribusiness, and the opening to a chemical assault on nature. If agricultural science were turned loose in poor countries, my friends reasoned, similar harms would be felt. Better for the poor to stick with low-tech agro-ecological farming systems that work by imitating rather than dominating nature. Better to use composted animal manure for fertilizer, rather than synthetic chemicals. And best to stay away from the dangerous new science of genetic engineering.

This postmodern resistance to agricultural science felt now in both North America and Europe makes considerable sense in rich countries, where science has already brought so much productivity to farming that little more seems needed. It becomes dangerous, however, when exported to countries in Africa where farmers remain trapped in poverty because they are starved for science. I show in this book that the turn in rich countries away from agricultural science has recently been exported inappropriately to Africa through a variety of international channels of influence, including foreign assistance programs, the United Nations, NGO advocacy campaigns, and commodity markets.

This export of rich-country tastes to the poor is most conspicuous in the case of genetically engineered agricultural crops, known as GMOs. But the issue goes way beyond genetic engineering, since a far more inclusive rich-country distaste for agricultural science is now being exported to Africa as well. Organizations from rich countries are coaching governments and farmers in Africa to stay away from modern science in farming, even though it is the same science that earlier reduced poverty in these wealthy countries. Instead of investing in new agricultural science, the poor are told to trust their own indigenous knowledge and to stick with the traditional crop and animal varieties currently in use by their own poor farmers. In effect, rich outsiders are telling African farmers it will be just as well for them to remain poor.

Many of the arguments put forward in this book go against dominant opinion in my own social circles. Some of my best friends would be troubled to learn that I endorse more chemical fertilizer use in Africa, rather than organic farming. Many of my colleagues and students would recoil if they knew I wanted to replace traditional African farming practices with an increased use of high-yielding crop varieties, or even worse, genetically engineered crops. In polite company the best social strategy for me has usually been not to talk about these ideas at all, but in this book I have decided to step out of that silence. Those who have helped me with this project should not be implicated in my conclusions; for these I take sole responsibility.

As I complete this work, I owe several important institutional debts of gratitude. First, I am eager to thank Wellesley College, my teaching home and primary provider of research support for the past three decades, which granted me the full sabbatical year I used to draft and edit the manuscript. Second, I thank Harvard University, where I maintain an ongoing appointment at the Weatherhead Center for International Affairs, and where I enjoyed a fellowship at the Belfer Center for Science and International Affairs during my 2006–07 leave year. Third, and with special enthusiasm, I thank the Rockefeller Foundation, which gave me and my wife Marianne the coveted opportunity to mix work with great pleasure for a month at their unsurpassed Study and Conference Center in Bellagio.

Less directly, much of the research for this project was carried out while I was working in Africa, or on Africa, for the International Food Policy Research Institute, the U.S. Agency for International Development, the African Center for Technology Studies, and the Common Market for Eastern and Southern Africa. I am grateful for the opportunity to be associated with all of these important and capable institutions.

Many individuals have been directly or indirectly helpful to me in the preparation of this work, some perhaps without realizing it. I want to make special mention of Marianne Banziger, Roger Beachy, Mopoko Bokanga, Joachim von Braun, Derek Byerlee, Joel Cohen, Christopher Darlington, Deborah Delmer, Chris Dowswell, Joseph DeVries, Jose Falck-Zepeda, Walter Falcon, Robert Falkner, Marnus Gouse, Michael Hall, Jeff Hill, Robert Horsch, Margaret Karembu, Jack Kyte, Josette Lewis, Isaac Minde, Stephen Mugo, Charles Mugoya, Larry Murdock, Chungu Mwila, James Ochanda, Steven Were Omamo, Ruth Oniang'o, Philip Paarlberg, Pilar Palacia, Rajul Pandya-Lorch, C. S. Prakash, Carl Pray, Gary Toenniessen, Laurian Unnevehr, David Wafula, and Judi Wakhungu.

Extra thanks go to those I have called upon several times for help and support in the course of this work, including Natalie DiNicola, Sakiko Fukuda-Parr, Lowell Hardin, Lawrence Kent, and Per Pinstrup-Andersen. Most of all, Calestous Juma, Michael Lipton, and Vernon Ruttan have been a source of consistent support and inspiration from the start, and much practical guidance at the end.

I am deeply grateful for the strong foreword to this book provided by Dr. Norman Borlaug and former president Jimmy Carter, both Nobel Peace Prize Laureates, both tireless champions for the betterment of Africa's rural poor, and both personal heroes of mine.

At Harvard University Press, my sincere thanks goes to Michael Fisher for approaching me to suggest a book of this kind, and then for his shrewd and well-informed insights in conceptualizing and developing the project. Kate Brick guided me through manuscript editing with experienced care.

For more intimate advice and personal support while executing this project I have relied almost entirely on my wife and partner, Marianne Perlak. She even agreed to come out of her recent retirement as an accomplished designer to provide the handsome design for this book. How many authors can be so lucky? To my wonderful Marianne, with thanks and love, I dedicate this work.

STARVED FOR SCIENCE

Why Are Africans Rejecting Biotechnology?

Consider a powerful new technology most people in rich countries do not need or even want: genetically engineered agricultural crops, otherwise known as genetically modified organisms (GMOs), or pejoratively as "Frankenfoods." The first generation of genetically modified (GM) crops appearing in the mid-1990s provided a tangible commercial benefit to farmers because this technology helped cut the costs of weed and insect control. But for food consumers the first generation of GM crops provided essentially no benefit at all. Genetically modified varieties of corn or soybeans did not taste any better, look any better, seem easier to prepare, or deliver greater nutritional value. Nor did they cost noticeably less. Absent any tangible benefit, a majority of citizens in rich countries—even in the United States—felt they could do without GM foods. In Europe the popular aversion to GMOs became so strong that stringent new regulations were placed on the technology, discouraging farmers from planting any GM crops.

It costs rich countries little when a new technology not needed or wanted by most citizens is driven off the market through stifling regulation. But what if the same stifling regulations then come to be adopted in poor countries with unmet farm-production and food-consumption needs? Many governments in Africa today, out of deference to the European example, have driven GM foods and crops out of their own markets by adopting European-style regulatory approaches toward the technology. European tastes regarding agricultural GMOs are not a good fit to the needs of Africa, given that two-thirds of all Africans are poor farmers in desperate need of new technologies to boost their crops' productivity. Keeping the new technology of genetically engineered crops away from these African farmers will come at a heavy and steadily increasing price.

This export to Africa of rich-country attitudes toward GMOs is part of a larger pattern. It isn't just about GMOs. Citizens in rich countries today have also grown to dislike most other applications of modern science to agriculture, including chemical fertilizers, pesticides, and specialized power machinery. Such technologies were the key to making farmland and labor more productive in today's rich countries, but they brought a highly specialized style of "factory farming" that holds little aesthetic or cultural appeal. Most nonfarmers in rich countries today do not like the idea of moving still further in the direction of industrial-style agriculture. Many now are choosing to pay whatever it costs to purchase foods that are "organically grown"—produced without the use of any modern synthetic chemical fertilizers or pesticide sprays, and without any GMOs. These citizens in rich countries regret the loss of traditional rural landscapes populated by small, diversified family farms, so they have begun calling for less agricultural science rather than more. They no longer want their tax dollars going for agricultural research and development (R&D), since they suspect this will only mean more power and profit for corporate agribusiness.

In a modern, globalized economy, what rich countries do and think at home seldom stays at home. The cultural and political turn against agricultural science that has taken place in the world's richest countries over the past several decades is now being exported, inappropriately, to the world's poorest countries, including those in Africa. When the governments of rich countries began reining in their investments in agricultural science at home, they simultaneously began withdrawing international assistance to agricultural science in poor countries abroad, with particularly damaging consequences for Africa. Asia and Latin America no longer need the help of rich countries in bringing appropriate applications of science to their farmers, but governments in Africa still do. Low productivity in farming is the trap that is currently keeping most Africans poor.

Africa's Farm Productivity Crisis

Because of nonproductive farming, Africa is now the only region on Earth where human poverty and hunger both continue to increase. Over the past fifteen years the number of Africans living on less than one dollar a day has increased 50 percent. Per capita income growth was negative between 1980 and 2000 (Sachs et al. 2004). The United Nations Development Programme concluded in 2004 that if current trends continue, Africa as a whole will not reach its 2015 Millennium Development Goal for human poverty reduction until the year 2147, more than a century behind schedule (UNDP 2004). With increasing poverty comes increasing malnutrition. Between 1991 and 2002 the number of undernourished people in the region increased from 169 million to 206 million. Nearly one-third of all men, women, and children in sub-Saharan Africa are currently undernourished, compared to just 17 percent for the developing world as a whole (FAO 2006).

It is especially perverse that the vast majority of Africa's malnourished poor are themselves food producers, working as smallholder farmers growing mostly food crops. It used to be suggested that Africa's food production crisis was being caused by an excessive emphasis on producing nonfood crops for export (Gakou 1987), but the vast majority of the region's total agricultural area and labor force remains dedicated to food-crop production for local use. In the semiarid regions of Africa where food production is failing most conspicuously, smallholder farmers routinely devote up to 90 percent of their cropped land only to the production of food grains for their own use or for local sale (Jayne 1994). The sorry truth is that food-crop production and export-crop production have *both* faltered badly in Africa in recent decades. According to one composite index of total agricultural production in Africa—food crop plus export crop—Africa's farms produced on a per capita basis 3 percent less in 2005 than in 2000, 12 percent less than in 1975, and 19 percent less than in 1970 (WRI 2006). In 1966–70, the African continent was still a net exporter of food, but by the late 1970s it was importing 4.4 million tons of food a year. By the mid-1980s it was importing 10 million tons of food a year. By 2002 sub-Saharan Africa was importing 19 million tons of food, in grain and grain equivalents, with more than 15 percent of these imports coming in the form of food aid (USDA 2004a).

This disastrous performance in Africa reflects, most of all, a failure to increase the productivity of human labor in agriculture. Roughly 70 percent of all African citizens still depend on agriculture for employment and income, and as long as the labor of these African farmers remains unproductive, they are doomed to remain poor. In Asia, where labor productivity in farming has been rising rapidly since the 1970s, hundreds of millions of rural dwellers have been able to escape poverty, but in Africa since the 1970s the productivity of labor in farming has remained low, and at times it has even declined. In Thailand between 1980 and 1997, average annual value added per farm worker increased by 50 percent, and in China by 100 percent, but in Africa it declined by 10 percent, from $418 to $379 (World Bank 2000). In one sampling of eleven of Africa's poorest countries—those where 35 percent or more of the population were malnourished between 1991 and 2003—annual value added per agricultural worker was declining in more than half. Farm labor productivity in these hungry countries averaged just $195 overall, well below one dollar a day (FAO 2006).

Africa's lagging agricultural performance is commonly attributed to a long list of factors not directly linked to the uptake of science on farms, factors such as landlocked geography, violent conflict, poor governance, HIV/AIDS, distortions in international markets, and climate change. Yet none of these factors can explain the depth, persistence, and pervasiveness of rural poverty in Africa. Landlocked states clearly have fewer options to pursue export-led growth, yet even in Africa's coastal states farmers remain nonproductive and poor. Violent conflict helps to explain cyclical hunger trends in Africa (nearly 80 percent of the increased incidence of hunger measured in the region between 1991 and 2002 took place in just five war-torn countries: Burundi, the Democratic Republic of the Congo, Eritrea, Liberia, and Sierra Leone), but overall levels of rural poverty and hunger in Africa remain high with or without violent conflict. Even in countries with no recent history of internal conflict such as Ghana, Kenya, and Tanzania (all coastal countries with excellent ports), farms remain unproductive and farm communities remain poor. Ghana, on the West African coast, has a good port, little or no history of ethnic fighting, a low prevalence of HIV/AIDS, and macroeconomic stability thanks to greatly improved governance over the past several decades, yet in Ghana annual value added per agricultural worker remains low at just $346—only one-fifth the level of El Salvador. Tanzania, another stable and peaceful coastal state with abundant access to international markets through the port of Dar es Salaam, remains stuck with an average annual value added per agricultural worker of only $290.

The poverty of African farmers cannot be explained by international market forces, since farmers in other developing countries who use the same markets are far less poor. International agricultural markets are undeniably distorted by the tariffs and subsidies of rich countries, yet the external barriers facing exports from Africa today are actually lower than those that faced the currently wealthy East Asian countries when they began their own period of high growth forty years ago (Yeats, Amjadi, and Reincke 1996). Even if all the world's remaining international agricultural market distortions were somehow eliminated—as they should be—the net gains for sub-Saharan Africa would be quite modest, probably equaling only about 1 percent of GDP (Tokarick 2003). More advanced and productive farmers in Southeast Asia, China, and Brazil would gain far more from global market liberalization than African farmers; in a free world market it is the most productive, not the poorest, who win. As the International Food Policy Research Institute (IFPRI) has concluded, "In the absence of productivity growth in the [African farming] sector, the agricultural supply response to reduced impediments to trade would be weak" (Omamo et al. 2006, p. 35).

Nor are HIV/AIDS or climate change convincing explanations for rural poverty in Africa. These problems are now making a bad situation worse, but they cannot possibly be the cause of low growth in farm productivity in the region because that low growth was already strongly visible in the 1980s, before any significant impact had yet been felt from these two more recent factors.

Farmers in Africa are poor because the productivity of their land and labor remains so low. Human labor in farming becomes more productive—and hence better paid—only when it has access to improved tools (such as metal plows), or when it begins to work with stronger and healthier animals (such as draft animals improved by modern breeding and veterinary medicine), or when it finds a more efficient way to replace the nutrients that cropping has taken out of the soil (for example, by adding chemical fertilizers), or when it grows crop varieties more likely to produce high yields or resist insects and plant disease (such as seeds improved by scientific plant breeders). Unfortunately, a preponderance of smallholder farmers in Africa have not made any of these gains. They still work their fields with hand hoes or crude wooden plows; they try to get their meat, milk, and draft power from animals stunted by poor health; they rely on traditional shifting cultivation practices despite population pressures on the land that shorten fallow times and mine nutrients from the soil; and they save and plant traditional crop seeds not yet significantly improved through scientific plant breeding. No matter how long or hard they work with these unimproved technologies, their productivity will remain constrained and their incomes will scarcely rise.

Within the constraints of their unimproved production technologies, farmers in Africa are remarkably efficient, yet they are still poor. They handle their tools, cropping systems, and animals with experienced judgment and exceptional skill. They allocate their labor efficiently among the complex list of daily tasks they must perform and nothing is wasted, yet profits from farming in Africa remain stuck at a low level. Crop production is so unrewarding a task for smallholder farmers in Africa that working-age men who do the heavy work of soil preparation at the beginning of the season often leave their families behind to tend the crops while they go looking for temporary employment in town.

Science as a Tool for Escaping Rural Poverty

Rural poverty of this kind was once the norm in Europe and North America as well. Three or four centuries ago ordinary farmers in Europe and America were poor (even those working on their own land) because they, too, were still using crude hand tools and working with unhealthy animals and unimproved, low-yielding crops. Their eventual escape from these impoverished rural conditions came when new discoveries in science were applied to farming, giving them steel plows, mechanical reapers, steam, gasoline and electrical power, improved breeds of cattle, better livestock feeds, inexpensive new chemical fertilizers, and finally hybrid seeds. This technology upgrade took different paths in different places. In North America land was abundant but labor relatively scarce, so labor-saving mechanical technologies proved to be of early value. In Europe labor was more abundant and the land frontier had long since closed, so chemical and biological technologies designed to make land more productive were naturally favored. As Chapter 1 will show, it was the availability of productive new technologies for farmers that allowed both Europe and North America, in the early and middle years of the twentieth century, to bring a final end to widespread rural poverty.

Then later in the twentieth century it became Asia's turn to find an appropriate package of technologies capable of reducing rural poverty through a boost in farm productivity. Asia's "Green Revolution" in the 1960s and 1970s was based on a biological science breakthrough: newly improved varieties of wheat and rice capable of producing much more grain in response to water and fertilizer inputs. These new seeds were developed originally by plant breeders in Mexico and the Philippines, then adapted to local conditions by international crop research centers and national scientists in South Asia and Southeast Asia, and distributed to farmers by national extension agents and nongovernmental organizations (NGOs). High yielding when adequately irrigated, fertilized, and protected against insects, these new seeds brought spectacular production gains just in time to support Asia's most rapid surge in population growth, helping to avert famine and permanent food aid dependence. Asia's annual rate of growth in rice output had been only 2.1 percent between 1955 and 1965, but thanks to the Green Revolution it increased over the next two decades to a significantly higher rate of 2.9 percent (Hayami and Ruttan 1985). India began planting the new wheat varieties in 1964, and by 1970 production had nearly doubled. India's rice production then doubled as well in the states of Punjab and Haryana between 1971 and 1976.

These farm productivity gains were strongly heralded at the time as a welcome escape from the looming threat of a massive Asian famine. In 1967 William and Paul Paddock had written a widely credited best-selling book (William was an agronomist and Paul was a former State Department Foreign Service officer) entitled *Famine 1975!*, projecting that India would never be able to feed its rapidly growing population. The Paddocks were so sure of this prediction they even advised against giving more food aid to India, as aid would only keep people alive long enough to parent still more children, who would then be certain to starve in even larger numbers in the future (Paddock and Paddock 1967). Within only a few years the Green Revolution reversed such expectations, and by 1975 India's farm-production gains were so great that the government found itself able to terminate all food aid deliveries on its own initiative. Instead of suffering a famine, India was able in 1975 to celebrate the attainment of national food independence.

Asia's science-based Green Revolution also disproved the fears of pessimists and critics by averting an environmental calamity. In 1964 India had been producing 12 million tons of wheat on 14 million hectares (ha) of land (one hectare is roughly 2.5 acres), but thanks to the high yields of the Green Revolution it was able by 1993 to increase its wheat production nearly four-fold while increasing its cropped wheat area by only 60 percent. To have produced this much wheat before the Green Revolution so successfully, increased yields would have required bringing much more land under the plow. In effect, the Green Revolution allowed India to meet its rapidly growing food needs without having to plow an additional 36 million hectares of cropland. M. S. Swaminathan, the charismatic Indian plant scientist who led the local crop-breeding effort, concluded: "Thanks to plant breeding, a tremendous onslaught on fragile lands and forest margins has been avoided" (Swaminathan 1994).

Most important for purposes here, the technologies of the Green Revolution in South Asia also helped reduce rural poverty. Since the improved seeds were a biological technology rather than a mechanical technology, they increased the productivity of land in a scale-neutral fashion, meaning small as well as large farms could share in the benefit. Small farmers could take up the technology so long as they had access to lands that were irrigated or well watered by rainfall, access to credit for the purchase of chemical fertilizers and insecticides, and access to adequate road and transport infrastructures to deliver their harvest to a commercial market, conditions generally met in most of the irrigated regions of South Asia (Ruttan 2004). Even the landless rural poor stood to benefit when the technology was taken up, because the added grain gave rural laborers more to harvest, transport, and process, which pushed up rural wages. One survey in southern India concluded that between 1973 and 1994 the average real income of small farmers rose 90 percent, while the incomes of the landless actually increased 125 percent (World Bank 2001).

A similar technology upgrade also helped pull farmers in China out of deep poverty at the end of the twentieth century. Between 1975 and 1990, new rice technologies such as the hybrid varieties developed by China's own scientists contributed over half (60 percent) to that nation's overall increase in average rice yields, and these gains helped bring more than 200 million people in rural China out of poverty between 1978 and 1999, the single largest mass escape from human poverty ever recorded over a two-decade period (Huang and Rozelle 1996, Chen 2000). In both China and India, investments in agricultural science were essential to reducing poverty. Models of the Chinese economy suggest that investments in agricultural research and development were more important than any other investments in stimulating overall agricultural GDP growth, and second only to education investments in reducing total numbers of people in poverty (Fan, Zhang, and Zhang 2002). Likewise in India, agricultural R&D was the best investment made by the state for the purpose of promoting agricultural growth, and second only to rural road investments for lifting people out of poverty (Fan, Hazell, and Thorat 2000).

Science investments should be employed today to promote a comparable escape from hunger and poverty for farmers in Africa. In 2004 Kofi Annan, then secretary-general of the United Nations, delivered this message publicly to a summit of African leaders in Addis Ababa:

> We are here together to discuss one of the most serious problems on Earth: the plague of hunger that has blighted hundreds of millions of

> African lives—and will continue to do so unless we act with greater purpose and urgency. The numbers are all too familiar. Nearly a third of all men, women, and children in sub-Saharan Africa are severely undernourished. Africa is the only continent where child malnutrition is getting worse rather than better . . . In Asia, Latin America, and the Middle East, a green revolution tripled food productivity and helped lift hundreds of millions of people out of hunger. Africa has not yet had a green revolution of its own. This is partly because the scientific advances that worked so well elsewhere are not directly applicable to Africa . . . That is why I have challenged the world's scientists and scholars to give their ideas, innovations, and intensity, and called on them to rally round the cause of food security and agricultural development in Africa . . . Let us generate a uniquely African green revolution—a revolution that is long overdue, a revolution that will help the continent in its quest for dignity and peace. (Annan 2004)

Africa lags in farm productivity not because science has failed, but because Africa's political leaders—the same leaders Kofi Annan was addressing—have largely failed to invest in science. Between 1981 and 2000, while per capita public spending on agricultural science was increasing by 30 percent in the developing world as a whole, it actually fell by 27 percent in Africa (Pardey et al. 2006). On top of this, in many African countries, policymakers have recently been denying their own farmers access to agricultural science through official disapprovals or stifling regulations placed on modern agricultural biotechnology.

Genetically Engineered Crops

Asia's successful Green Revolution in the 1960s and 1970s was based on new seeds developed through conventional plant breeding techniques, including hybridization. More recently, crop scientists have learned to improve plants through an additional technique known as genetic engineering, or recombinant DNA (rDNA) science. This method does not rely on the pollination of plant flowers; it allows individual genes with desired traits to be moved directly from one organism into the living DNA of another. Genetic engineering was first accomplished in the laboratory in 1973, and soon found commercial applications in medicine. In 1982 the Food and Drug Administration in the United States approved the use of human insulin produced by a genetically engineered bacterium. Genetically engineered animal vaccines came next, followed by genetically engineered agricultural crops, first approved for commercial use in the mid-1990s.

Genetic engineering is just one of several tools in the crop science kit. The traditional approach is conventional selection breeding, cross-pollinating plants that are sexually compatible and then selecting among the offspring those with the most desirable traits. Early in the twentieth century, scientists began to develop more controlled and powerful forms of cross-pollination, such as hybrid crosses between carefully inbred parent lines. Advances in the science of genomics eventually made possible an even more precise and powerful conventional breeding method, known as marker-assisted selection (MAS). This technique is based on knowing the molecular markers of desired genetic traits and then using tissue screening to determine the presence or absence of such traits even before a plant matures, thus speeding the breeding process and making it more precise.

Genetic engineering goes one step further by allowing crop scientists to bring in genetic traits from species that are beyond the normal reproductive range of the plant being improved. For example, a gene from a bacterium can be inserted into the cells of a crop plant to provide resistance to some kinds of caterpillars. Once a desired trait has been inserted in this fashion into the living DNA of a host plant, it becomes inheritable when that plant reproduces naturally. An unfounded rumor was spread in the 1990s by critics of the technology that GM seeds could not be reproduced by farmers because they had been sterilized by "terminator genes." The only basis for this rumor was the existence of a patented technology that might have been able to produce sterile seeds, but the technology only existed on paper; it has never been introduced into any of the GM plants currently on the market, and the largest developer of GM seeds, Monsanto, has publicly pledged it will not do so (Monsanto 1999). The GM crops on the market today reproduce just as easily as their non-GM counterparts, and in a number of countries, including Brazil, India, and China, the ease with which they can be replicated by farmers has actually been a key factor in accelerating their spread.

When genetic engineering techniques were first developed in the 1970s, private companies quickly saw their commercial potential for improving agricultural plants. Crop improvement is big business, with annual global seed sales to commercial farms now valued at more than $17 billion. Most of these sales are made to prosperous farmers in countries within the temperate zone, so private seed and biotechnology companies naturally began by developing genetically engineered crop varieties intended specifically for these customers. The first generation of transgenic crops were cotton and corn plants engineered to resist insect damage (reducing the need for costly insecticide sprays) and soybean plants engineered to resist a specific herbicide (in a manner that lowered the cost of weed control). As these new technologies began working their way through the private corporate research pipeline in the 1980s, regulatory authorities in most of the wealthy countries where farmers were expected to buy the new seeds, including the United States, Canada, European countries, Japan, Australia, Argentina, and South Africa, all began setting in place formal risk-assessment systems to screen the new genetically engineered crops on a case-by-case basis for evidence of harm either to human health or to the environment. Based on successful risk assessments, government regulators in all of these countries began by 1995–96 to approve a first generation of genetically engineered crops for commercial planting and human consumption. Regulators in the United States approved genetically engineered varieties of soybean, corn (maize), cotton, canola, tomato, and potato. Canadian regulators gave a nearly parallel set of approvals at almost the same early date. The European Union in 1995–96 approved the planting or consumption of genetically engineered canola, soybean, and maize. Japan approved soybean and tomato. Argentina approved soybean and maize. Australia approved cotton and canola. Mexico in 1995–96 approved soybean, canola, potato, and tomato (James 2006).

Africa's Surprising Rejection of GMOs

When these more prosperous countries began approving GM crops in 1995–96, nobody expected the technology would be rejected in Africa.

There were worries that poor farmers in Africa wanting to plant GMOs would have trouble getting their hands on the valuable new technology due to cost or patent restrictions, but nobody suspected it would eventually be African governments that would block access by imposing stifling regulations on GM foods and crops.

Africa's inclination to reject agricultural GMOs originally surfaced in the context of an international negotiation launched in 1996 under the United Nations Convention on Biological Diversity (CBD), the negotiation of a "biosafety protocol" to ensure that international trade in living genetically engineered crops or seeds (called living GMOs, or LMOs) did nothing to compromise the safety of the biological environment. African governments were thus introduced to GMOs in the narrow context of possible environmental risks. Africa's response in this negotiation was led by the environment minister of Ethiopia, who had become persuaded—in part by environmental activists from Europe—that GMOs should be regulated in much the same way an earlier 1989 agreement had regulated international trade in hazardous wastes. No new risks to the environment had yet been documented, yet by the time these international negotiations concluded in 2000, with a new agreement called the Cartagena Protocol on Biosafety, most African governments had come to believe agricultural GMOs were inherently risky.

Even some African governments that had been willing to sponsor research on genetically engineered crops earlier in the 1990s then began getting cold feet. Egypt had created the Agricultural Genetic Engineering Research Institute (AGERI) in 1989, with plans to use genetic engineering to improve a number of Egyptian crops, including potatoes, maize, and tomatoes. Yet despite successful field trials in 1997 with a GM potato resistant to insect damage, Egypt never approved the potato for commercial production. No evidence of food safety or biosafety risk had turned up, but fears had developed that the GM potatoes might be rejected by importers in Europe (USDA 2006b).

The government of Kenya also got cold feet. In 1991 Kenya's Agricultural Research Institute (KARI) had been approached by the U.S. Agency for International Development (USAID) with an offer to develop a GM sweet potato, and KARI agreed. But it subsequently took six years for Kenya's National Council for Science and Technology

(NCST) to issue regulations and guidelines to govern the safe handling of GMOs in the country, and it took another two years for Kenya's National Biosafety Committee to approve an initial import of the materials for research purposes. Further delays then slowed the field trials for this disease-resistant sweet potato, and despite an absence of evidence of risk, it still has not been approved for commercial planting. Kenya subsequently allowed trials of GM maize and cotton, once again with no recorded evidence of biosafety harm, but no approvals have yet been given for commercial release.

The growing governmental resistance in Africa to agricultural GMOs came to full bloom in 2002 when controversies arose over whether or not to import GM maize from the United States as food aid. Prior to 2002, shipments of unmilled GM maize from the United States had been accepted in Africa without controversy. This genetically engineered maize had been widely planted (and consumed) in both the United States and Canada since 1996, and as one Kenyan official had said in 2000, "Our confidence was established in the fact that if Americans are eating it, it should be safe for our starving people" (Mugabe et al. 2000).

All this changed in 2001–2002 when a severe drought struck Southern Africa, leaving 15 million people across seven countries facing serious food deficits. In May 2002, just as this crisis was intensifying, the government of Zimbabwe decided to turn away a 10,000-ton shipment of unmilled U.S. corn, expressing an official concern that some of the GM kernels might be planted rather than eaten. This forced the shipment to be diverted to other countries in the region also facing food shortages, including Mozambique, Malawi, and Zambia. Zambia, however, took even stronger action. In August 2002 the vice president of Zambia provisionally turned down all imports of GM maize, even though nearly 3 million of his citizens faced a pressing need. Zambian leaders had been importing GM maize as food aid for a number of years, but now they were refusing it, even in an emergency. At an inflammatory public meeting the leader of a local women's NGO had told them in no uncertain terms that even hungry Zambians did not want this type of American food aid (Phiri 2002). Zambian president Levy Mwanawasa later commented, "Simply because my people are hungry, that is no justification to give them poison, to give them food that is intrinsically dangerous to their health" (BBC 2002).

The head of the local United Nations World Food Programme (WFP) office in Lusaka implored the Zambian government to change its policy, but without success. WFP was thus obliged to begin removing from Zambia the GM food aid supplies it had delivered earlier, and in January 2003 this led to an embarrassing incident when a mob of villagers in the town of Sizanongwe, 300 kilometers from the capital, overpowered an armed guard and looted several thousand bags of the food aid before it could be removed (IRIN News 2003).

Additional African states took less drastic but equally telling measures against GM food aid. Malawi and Mozambique decided to place restrictions on GM food aid similar to those Zimbabwe adopted, requiring that whole kernels of maize be milled prior to distribution so as to prevent planting. As a result of these demands, WFP needed to reroute and even reverse a number of food aid deliveries to the region. Large shipments of GM maize were stranded at ports of entry, and WFP was forced to make emergency arrangements to mill large quantities of GM maize in South Africa, reopening previously mothballed mills. The process of milling added $25 per ton to the cost of delivering the food aid and resulted in delivery delays (Bennett 2003). Nevertheless, at a meeting in Dar es Salaam in 2003, the fourteen Southern African Development Community (SADC) countries adopted official guidelines that endorsed the milling of GM maize prior to its distribution as food aid, plus local or regional sourcing of as much food aid maize as possible, to avoid maize with GM content from the United States or Argentina.

This new pattern of rejecting GMOs spread to Angola in April 2004 when that government began refusing unmilled GM maize from the United States as food aid, even though WFP was already finding it hard to deliver feeding rations for refugee repatriation in the country due to funding shortfalls. Later that year even the government of Sudan took time out from its genocidal suppression of a rebellion in Darfur to issue a memorandum requiring that all food aid brought into the country should be certified as free of any GM ingredients. As of 2006, Angola, Malawi, Mozambique, Namibia, Nigeria, Zimbabwe, and Sudan had all officially rejected food aid shipments that might contain unmilled GM grains, while others, including Zambia, Ghana, and Benin, had announced explicit bans on the import of all GM foods and crops.

Governments in Africa have become so wary of agricultural GMOs that only one state on the continent, the Republic of South Africa, has yet made it legal for ordinary farmers to plant such crops. Africa needs new farming science more than any other region, yet it has been the least eager, so far, to embrace modern agricultural biotechnology.

External Sources of Africa's Behavior

Africa's rising rejection of genetically engineered crops despite an acute need for food aid calls for some explanation. These are the same crops ordinary Americans have been growing and consuming since 1995 without any documented mishap. Is this another sad example of Africa's readiness to accept quack science, as in the case of South African president Thabo Mbeki's damaging rejection of the science demonstrating HIV to be the cause of AIDS? Or does this perhaps reveal something about Africa's cultural insularity from the West, similar to the persistence of polygamy in many African communities today? In fact, Africa's rejection of genetically engineered crops today is far more western than it is African. Governments in Africa did not begin to get cold feet about GM crops until they saw activists and consumers in rich countries—particularly in Europe—rejecting the technology.

The modern political rejection of GM foods and crops had its beginnings not in Africa but in Europe, where the technology lost favor in the wake of an essentially unrelated 1996 food-safety scandal over meat contaminated by bovine spongiform encephalopathy (BSE), better known as "mad cow disease." This disease had been damaging the health of beef cattle in the United Kingdom since the 1980s, but government regulators (partly compromised by their close ties to the beef industry) had assured consumers that meat from the diseased animals was safe to eat. Then in 1996 the British government was obliged to reveal that ten cases of an incurable and fatal human disease called Creutzfeld-Jakob disease (CJD) were indeed linked to the consumption of BSE-contaminated meat. It was precisely at this moment in the spring of 1996, by an unhappy coincidence, that EU regulatory authorities gave their first official approval for the import and consumption of a GM food, an herbicide-tolerant (Roundup Ready) soybean developed by Monsanto in the United States.

Activist NGOs in Europe such as Greenpeace, Friends of the Earth, and the European Consumers' Organisation (BEUC) could see no consumer benefit in GM foods that might justify even a hypothetical safety risk, so they began warning citizens away from GM foods and crops simply on precautionary grounds. Efforts by European regulators to reassure consumers had no impact, since the credibility of these regulators had been destroyed by the erroneous safety assurances earlier given in the case of BSE. European supermarket chains began removing known GM products from their shelves to avoid being targeted by activist demonstrators, yet anxieties grew, and in June 1997 the European Union decided under citizen and activist pressure to require that all GM food sold in Europe carry an identifying label. Instead of reassuring consumers, this step only reinforced the growing impression that GM foods must indeed be dangerous. In 1998 political anxieties about GM foods and crops in Europe grew so intense as to oblige EU regulators to place an unofficial moratorium on any new case-by-case approvals of GM crops.

This political and regulatory backlash against GM foods and crops in Europe took place despite an absence of any credible scientific evidence linking GM products approved by regulators to any new risks to human health or the environment. However, since consumers saw no benefit in the new technology and since official regulators were so little trusted, it became easy for activists to stigmatize GM products on precautionary grounds.

There never was a comparable regulatory backlash against GM foods and crops in the United States for a number of reasons. The United States experienced nothing in the mid-1990s comparable to Europe's BSE crisis, and America's tradition of market capitalism remains distinctly less prone to regulatory intrusion than the "coordinated market capitalism" tradition of Europe (Hall and Soskice 2001). Also, the American legal system tends to use civil litigation after the fact rather than pre-emptive regulation before the fact to ensure consumer and environmental safety. And finally, America's two-party political system gives less space for Green Party candidates to gain election and then join governing coalitions to advocate against GMOs. Although regulatory outcomes have diverged for all these reasons, at the level of social opinion the similarities in response to GMOs in these two wealthy regions actually outweigh the differences. Environmental and consumer advocacy groups in the United States are nearly as firm and vocal in their opposition to GM foods and crops as counterpart groups in Europe.

It's Not about Genetic Engineering

If genetic engineering were really the issue, citizens in rich countries would be just as skeptical today about the use of GMOs in medicine as in agriculture, yet citizens in both the United States and Europe welcome the recombinant drugs made from GMOs that now constitute approximately 25 percent of all new drugs approved for the market in rich countries. Wealthy societies provide strong support for state-of-the-art medical science (whether it involves genetic engineering or not) because their desire for improved health and greater longevity is insatiable. In contrast, added agricultural productivity is something rich countries do not so urgently need or want. So it is this absence of a compelling benefit, not the presence of a possible risk, that most shapes opinion toward agricultural GMOs in rich countries. Such a conclusion challenges the work of Ulrich Beck, who has argued that wealthy societies are "risk societies," organized largely for the purpose of cutting risks (Beck 1992). Beck and others have asserted that as wealth piles up, the prospect of loss becomes more salient than any prospect for added benefit, reducing social enthusiasm for the risks that might come from further modernization. The comparison between medical GMOs and agricultural GMOs shows to the contrary that risks are not that important to rich societies, so long as tangible benefits are available. When tangible benefits are not available, technologies can be rejected even in the absence of any documented risk.

The dislike of agricultural GMOs in rich countries—including a dislike by many citizens in the United States—reflects more than just an absence of direct consumer benefits. Chapter 2 will show it is part of a larger turn against any new application of science to agriculture. Rich countries have seen the productivity of their own farms increase exponentially through applications of science, so much that they do not really want any more. This turn against new agricultural science is an affordable attitude in rich countries, but it becomes dangerous if exported to science-starved poor countries where farmers are not yet productive or prosperous. Chapter 3 will show that when governments in rich countries began downgrading their own public investments in new agricultural science at home in the 1980s, they simultaneously withdrew support for investments in agricultural science in poor countries abroad, inducing aid-dependent governments of Africa to follow suit, thus worsening the plight of African farmers.

Chapter 4 will return to the specific issue of GM crops, showing this to be an extreme case of rich countries, especially European countries, discouraging Africa from embracing modern agricultural science. Policy elites in Africa are now being prompted not to use agricultural GMOs through the influence of foreign assistance programs operated by European states, by intergovernmental organizations and international NGOs funded and managed from rich countries (again, mostly from Europe), and also through international commodity markets where rich countries—especially European countries—dominate as big customers. Because of such external influences, Africa's rejection of agricultural GMOs is actually more European than it is African.

In Chapter 5, I test the strength of these mechanisms of external influence by examining a dramatic new application of genetic engineering that should be easy to support because it is so precisely suited to the needs of poor farmers in Africa: GM crops engineered to produce grain under drought-stress conditions. Even in the case of drought-tolerant crops, the international community has yet to mobilize adequate resources to make this new technology available to the poor in Africa.

In the conclusion I look inside Africa to ask more directly why governments there have found it so easy to sacrifice the interests of their own poor farmers by underinvesting in agricultural science overall since the 1980s, and now by resisting GM crops in particular. I find an explanation in the strong international dependence (commercial, financial, and cultural) that Africa's urban political leadership class still feels toward Europe. Through what can be described as "an imperialism of rich tastes," urban political elites in Europe transplant their policy preferences regarding GMOs (and agricultural science in general) into the thinking of Africa's urban political elite. In the end it is not Africa's science-starved farmers that are rejecting biotechnology; it is Africa's urbanized governing elites that are doing so, prompted by their continued deference to urbanized elites in Europe.

1

Why Rich Countries Dislike Agricultural GMOs

Applications of genetic engineering to food and agriculture are unpopular in all wealthy countries, among many citizens in the United States as well as Europe. Different regulatory systems have emerged in America (a permissive system) versus Europe (a highly precautionary system), but these regulatory differences mask an underlying similarity: most consumers on both sides of the Atlantic dislike genetically modified crops and will avoid GM foods if given an easy means of doing so. Here I explain the origins of this important underlying pattern of opinion in wealthy countries. New risks to human health or to the environment are not the explanation, since no such risks have yet been documented. Nor is a dislike of genetic engineering the explanation, since the same citizens who dislike GMOs in agriculture welcome GMOs in medicine. The comparison to medicine also suggests this is not a popular dislike of multinational corporations, or of their private patenting of science, or of the high cost of their products, or even of their corrupting influence over regulators, since all of these disagreeable elements are also present in the case of GM drugs, which consumers do not reject.

From the perspective of social acceptance, there are two dimensions along which GM crops and foods differ significantly from GM drugs: environmental release and involuntary consumption. Genetically modified crops grow in an open environment whereas recombinant GM drugs are produced under strict containment in medical laboratories. Genetically modified foods are also sold into commercial food markets, often without segregation or labeling, whereas GM drugs are strictly segregated, labeled, and individually prescribed by physicians. Both these factors emerge as more plausible explanations for why citizens in rich countries like GM drugs but not GM foods and crops. Yet the single best explanation is not linked in any way to risk, cost, or even control over exposure. The best explanation for why rich countries dislike GM foods and crops is the absence for most citizens of a direct and noticeable benefit.

The first generation of GM foods and crops on the market provided significant cost savings for farmers, and profits for seed companies and patent holders, but in today's wealthy postagricultural (indeed, postindustrial) societies these beneficiary groups are numerically small. Only 2–5 percent of citizens in rich countries work today in any kind of farming. For the 99 percent or more of citizens in rich countries who do not plant corn, soybeans, or cotton, or sell the seeds of these crops, or own the patents on those seeds, the first applications of genetic engineering to agriculture provided no direct or tangible benefit. Recombinant GM drugs, in sharp contrast, provided potential lifesaving benefits to citizens from every demographic category. It is not risk or cost or uncontrolled exposure that drives citizen perception of the science behind GMOs in rich countries, but instead a calculation of likely benefits.

Citizen Opinion toward GM Foods and Crops in Rich Countries

The United States and Europe have made different regulatory choices toward GM foods and crops, yet on both sides of the Atlantic most ordinary citizens tend to dislike this new technology. Ordinary Americans are not known for high standards or refined taste when it comes to food, yet when asked about eating genetically modified foods, a plurality say they would rather not. In response to a Pew Initiative survey in 2005, half of a representative sample of Americans even said they would oppose the introduction of genetically modified foods into the U.S. food supply, with 33 percent saying they would oppose GM foods strongly (Pew Initiative 2005). This is a response that reveals considerable ignorance along with opposition, because food products with GM ingredients (mostly corn and soy ingredients) have been pervasive in American supermarkets since 1996.

As of 2006 an estimated 61 percent of all corn grown in the United

States and 89 percent of all soybeans were GM varieties. In the United States these GM crops have been marketed and processed with no mandatory segregation from conventional corn and soybeans, and as a result roughly 70 percent of all supermarket products in the United States have at least some GM content. The only Americans today not consuming at least some GM foods are the tiny minority who consume only organically grown foods. Yet the Pew survey found that in 2005 only 25 percent of U.S. respondents believed they had ever eaten genetically modified foods. An earlier 2005 survey by the International Food Information Council found only one-third of consumers in the United States were aware that GM foods were being sold in stores (IFIC 2005).

One reason for this consumer ignorance in the United States is an absence of labeling requirements for GMOs. The U.S. Food and Drug Administration (FDA) holds to a view that labeling of GM foods should only be mandatory if they are shown to differ significantly in composition from their conventional counterparts in some way that might pose a risk to the consumer—such as through the presence of an allergen, a changed level of a major dietary nutrient, an increased level of toxins, or a change in the expected storage or preparation characteristics of the food (Miller and Conko 2004). So far, none of the GM foods on the market has been shown to differ enough from the conventional counterpart to require such added labeling. Yet when Americans are asked directly if they would like to see all GM foods labeled, 94 percent say yes (Hallman et al. 2003). Lawsuits and lobbying by consumer advocate groups have so far failed to force a change in the FDA's position, so GM foods continue to appear on store shelves in the United States unlabeled. In the European Union labeling became mandatory in 1997, but by then most retail stores had already decided voluntarily not to stock any GM products so as to avoid boycott campaigns from activists. Labeling outcomes in Europe and the United States are thus quite different, but with neither measuring up to an ideal of "informed choice." In the United States consumers have a choice between GMO and non-GMO but no information, while in Europe consumers are guaranteed information but with no choice, since only non-GM products can be found on the shelf.

When American consumers are questioned directly about GM foods, the responses are surprisingly negative. A 2003 survey by the Food Policy Institute at Rutgers found that fewer than half (45 percent) of Americans felt it was safe to consume GMOs, while more than half (54 percent) felt that "GM food threatens the natural order of things," and 62 percent felt that "serious accidents involving GM foods are bound to happen" (Hallman et al. 2003). Two-thirds of all Americans assert they would oppose importing GM foods into the country, a surprising view from residents of the world's largest producer and exporter of GMOs (Pew Initiative 2005). A 2006 Cornell University survey found that trust in agricultural biotechnology in the United States was not only low but actually declining between 2003 and 2005 (Peterson 2006).

Food companies and fast-food franchises in America have sensed that consumers might turn away from any foods known to be GMOs and have therefore been shy about using them. Genetically engineered potatoes resistant to viruses and beetles were successfully planted in the United States between 1995 and 1999, but fast-food chains such as McDonald's and Burger King and food companies such as McCain Foods and Procter & Gamble, maker of Pringles, thought it commercially wise to demand from processors only non-GM potatoes for their chips and fries. Potato processors consequently began contracting with farmers for only non-GM potatoes, and eventually in 2001 Monsanto voluntarily withdrew its genetically modified NewLeaf potatoes from the market. Similarly in 2000, Pepsico's Frito-Lay company decided to require non-GM corn from its contract growers for manufacturing its snack products. Del Monte and other food companies have rejected GM sweet corn. Genetically modified popcorn has been approved for consumption by regulators in the United States, but U.S. popcorn manufacturers have decided voluntarily not to use it, fearing consumer resistance (CFS 2006).

Citizen dislike of GM foods is stronger in Europe than in the United States, but not by any order of magnitude. A comparative survey of attitudes in 2001 found the ratio of Americans who said they "supported" the use of biotechnology in crop production, versus those that did not, was essentially balanced, with 32 percent in support versus 31 percent opposed, the rest saying they did not know. When the same survey was administered in the United Kingdom, the balance between supporters and opponents in the United Kingdom was 38 percent in support versus 46 percent opposed, less favorable than in the United States but not dramatically so (Moon and Balasubramanian 2001).

Beyond rich countries, citizen opinion about agricultural GMOs tends to be somewhat less hostile. In 2000 an internationally comparable opinion survey asked 35,000 respondents in 35 different countries if they agreed with the (admittedly complex) statement, "The benefits of using biotechnology to create genetically modified food crops that do not require chemical pesticides are greater than the risk." Only 22 percent of citizens in France and Greece and only one-third of Japanese respondents agreed the benefits were greater, but in China and India the level of citizen agreement with this statement was above 65 percent, even higher than the level in the United States. Citizen approval of GMOs was actually highest in Cuba (79 percent) and Indonesia (81 percent) (Hoban 2004). A 2006 study by the global market research company Synovate found that, among consumers who were aware of GMOs, 89 percent of Greeks believed they may be harmful, yet only one-third of citizens in South Africa felt the same way. It is unsurprising that citizens in countries who have less reason to be satisfied with the status quo are more open to the use of new food and farm technologies.

Citizen views toward GMOs can of course differ dramatically person by person and state by state. In the United States disapproval is strongest among people over sixty-four, among women, and among people with low levels of education. An identical pattern emerges in Europe (Gaskell et al. 2000). Americans with postgraduate degrees are among those most likely to approve of GMOs. Approval also correlates with high income, but not independent from educational attainment. Different eating preferences are of course also a key variable. Americans who have adopted restrictive dietary practices (vegetarian, vegan, Kosher) and those who say they favor "natural" or "healthful" or "organic" foods are among those most likely to oppose GMOs (Hallman et al. 2003). In some cases, citizen opposition to GMOs in the United States has been strong enough to dictate local political outcomes. In 2004 when an organization of organic farmers championed a ballot initiative in California's Mendocino County to ban genetically engineered crops and animals, the measure passed by a slim margin (56 percent in favor and 44 percent opposed). By 2006, three other California counties—Marin, Santa Cruz, and Trinity—had adopted similar measures.

Social skepticism about GM foods and crops in America has recently been strong enough to slow the approval of new products by government regulators. Between 1995 and 1999, when GMOs were new, a total of forty-seven such crops were reviewed by the FDA and then granted deregulated status. No subsequent damage to human health or the environment from any of these first approved products has ever been documented, yet subsequently, from 2000 to 2004, the number of new GM crops completing the FDA review process fell to just fifteen, and only thirteen of these were commercialized. Also during this second period, the only new GM varieties commercialized were crops previously commercialized in an engineered form (corn, cotton, canola, sugar beet) and containing transgenes either identical or similar to those earlier approved. Also, regulators had begun taking twice as much time, on average, to reach their decisions (Jaffe 2005, 2006). Even in America, then, agricultural GMOs have remained under a cloud of latent but widespread citizen suspicion and have encountered increasing regulatory caution.

Citizen Skepticism Despite No Evidence of New Risks

Explaining this pattern of citizen skepticism and regulatory caution toward GMOs in rich countries is a challenge, given the absence so far of any scientific evidence of new risks from the technology. Risk is conventionally defined as the probability (high or low) of an unwanted event. Citizen opposition to GM foods and crops is often expressed in terms of risk avoidance, but the presence of new risks has yet to be detected and documented by scientific authorities. In both Europe and the United States scientific authorities have repeatedly asserted there is not yet any credible evidence of new risks to human health or the environment from any of the GM foods or crops approved by regulators and placed on the market so far.

In Europe this scientific consensus was first summarized in 1999, when Britain's Nuffield Council on Bioethics published a report stating, "We have not been able to find any evidence of harm. We are satisfied that all products currently on the market have been rigorously screened by the regulatory authorities, that they continue to be monitored, and that no evidence of harm has been detected" (Nuffield Council 1999, pp. 126–127). Two years later the Research Directorate General of the European Union reached a similar conclusion, releasing a summary of eighty-one separate scientific studies conducted over a fifteen-year period (all financed by the European Union rather than private industry) aimed at determining whether GM products were unsafe, insufficiently tested, or underregulated (Kessler and Economidis, eds. 2001). The EU Research Directorate concluded from this study, "Research on GM plants and derived products so far developed and marketed, following usual risk assessment procedures, has not shown any new risks on human health or the environment." (EU Research Directorate 2001).

National academies of science in Europe began drawing this same conclusion one year later. In December 2002, the French Academy of Sciences stated that "all the criticisms against GMOs can be set aside based for the most part on strictly scientific criteria" (French Academy of Sciences 2002, p. xxxviii). At the same time the French Academy of Medicine announced it had found no evidence of health problems in the countries where GMOs had been widely eaten for several years (French Academy of Medicine 2002). In the United Kingdom in May 2003, the Royal Society presented to a government-sponsored review two submissions that found no credible evidence GM foods were more harmful than non-GM foods, and Professor Patrick Bateson, the vice-president and biological secretary of the Royal Society, expressed irritation at the undocumented assertions of risk that continued to come from opponents of GMOs: "We conducted a major review of the evidence about GM plants and human health last year, and we have not seen any evidence since then that changes our original conclusions. If credible evidence does exist that GM foods are more harmful to people than non-GM foods, we should like to know why it has not been made public" (Royal Society 2003). In March 2004, the British Medical Association (BMA), which had earlier withheld judgment, endorsed these Royal Society conclusions (BMA 2004). In September 2004, the Union of the German Academies of Science and Humanities produced a report that concluded, "according to present scientific knowledge it is most unlikely that the consumption of the well characterized transgenic DNA from approved GMO food harbours any recognizable health risk" (Helt 2004, p. 4). This report added that food from insect-resistant GM maize was probably healthier than from non-GM maize due to lower average levels of the fungal toxins that insect damage can cause.

The most reputable international scientific and technical bodies have drawn equally benign conclusions. In March 2000, the Organisation for Economic Co-operation and Development (OECD) in Paris organized a conference with 400 expert participants from a variety of backgrounds. These experts announced their agreement that "No peer-reviewed scientific article has yet appeared which reports adverse effects on human health as a consequence of eating GM food" (OECD 2000, p. 2). In August 2002, the director-general of the World Health Organization (WHO) endorsed consumption of GM foods, saying, "WHO is not aware of scientifically documented cases in which the consumption of these foods has negative human health effects. These foods may therefore be eaten" (Mantell 2002).

Some who accept that GM foods are probably safe to eat still question their safety for other living things in the biological environment—their "biosafety." Since all farming disturbs and changes nature, it is difficult here to agree on what standard of nondisturbance to view as safe. For example, GM varieties of beet or rapeseed help farmers control weeds in the field (compared to conventional beet or rapeseed). As a result there are usually fewer weeds in the field and hence fewer weed seeds for some farmland birds to eat. Biosafety purists might view this as a harm to nature, yet to be consistent they would then have to oppose all weed removal from farm fields.

Using more conventional definitions of biosafety, the GM crops currently on the market have apparently not disturbed nature any more than conventional crops. A 2003 study conducted by scientists from New Zealand and the Netherlands published in *The Plant Journal* examined data collected worldwide up to that time, and the authors concluded from this data that the GM crops approved so far had been no more invasive or persistent than conventional crops and no more likely to lead to gene transfer. There was no evidence that GM crops had transferred to other organisms (including weeds) new advantages such as resistance to pests or diseases or tolerance to environmental stress (Connor, Glare, and Nap 2003). Later in 2003 the International Council for Science (ICSU) examined the findings of roughly fifty different scientific studies that had been published in 2002–03 and concluded, "there is no evidence of any deleterious environmental effects having occurred from the trait/species combinations currently available" (International Council for Science 2003, p. 3). In May 2004, the United Nations Food and Agriculture Organization (FAO) issued a 106-page report summarizing evidence that "to date, no verifiable untoward toxic or nutritionally deleterious effects resulting from the consumption of foods derived from genetically modified foods have been discovered anywhere in the world." On the matter of environmental safety, this FAO report found the environmental effects of the GM crops approved so far, including effects such as gene transfer to other crops and wild relatives, weediness, and unintended adverse effects on nontarget species (such as butterflies), had been similar to those that already existed from conventional agricultural crops (FAO 2004). Finally in 2007, a study done for the journal *Advanced Biochemical Engineering/Biotechnology* surveyed ten years of research published in peer-reviewed scientific journals, scientific books, reports from regions with extensive GM cultivation, and reports from international governmental organizations and found that "The data available so far provide no scientific evidence that the cultivation of the presently commercialized GM crops has caused environmental harm" (Sanvido, Romeis, and Bigler 2007).

A scientific consensus even exists that the GM crops on the market so far are producing some measurable benefits for the natural environment, due primarily to their ability to thrive with reduced sprayings of toxic chemicals and in some cases with reduced soil tillage. Studies of cotton production in Australia, China, South Africa, and the United States have shown reductions in insecticide spraying of 40 to 60 percent for GM cotton compared to conventional cotton crops (International Council for Science 2003). Reduced spraying of insecticide means less pollution of ground water and surface water and also less damage to nontarget species, such as the beneficial insects that live in and around the farm field. According to one 2005 calculation, the planting of GM crops up to that point had made possible a global reduction of 15 percent in the total volume of insecticides applied to cotton since 1996, and a reduction of 4 percent in the total volume of herbicides used on soybeans (Brookes and Barfoot 2005). Herbicide-tolerant GM soybeans can be grown not only with fewer, less toxic, and less persistent herbicide sprays, but also with less soil tillage, a factor that reduces erosion and the siltation of water bodies downstream (International Council for Science 2003). GMOs even help cut greenhouse gas emissions by reducing the burning of diesel fuel thanks to lower mechanical tillage requirements and a less frequent need for field applications of herbicides and insecticides. Over the period 1996–2004 a cumulative reduction in fuel use equal to 4.9 billion kilograms (kg) of carbon dioxide was made possible by farmers switching—mostly in North and South America—to GM crops. The adoption of "no till" and "reduced till" weed control systems (made possible with herbicide-tolerant GM crops) has also had an environmental benefit in the form of more carbon sequestration. In 2004, an additional 9.4 million kg of carbon dioxide was sequestered in the soil, thanks to reduced tillage made possible by GM crops (Brookes and Barfoot 2005).

It is quite unusual for any powerful new technology to perform during its first dozen years on the market with no documented evidence of any new harms or risks. Try to imagine the first dozen years of organ transplantation without anyone dying on the table, or the first dozen years of commercial aviation without a single plane crash. Skeptics try to deny this safety record by saying "absence of evidence is not the same thing as evidence of absence." It may not be *proof* of absence (since proving any negative is impossible), but if you look for new risks for a dozen years and fail to find any, that surely counts as evidence of something. Skeptics who hold out against GMOs because "proof" of safety is missing are measuring the technology by an unreasonable standard. If employed consistently, such an approach would oblige us to reject every new technology that comes along, and in fact every old technology as well.

Some critics try to argue that the strong safety record of GMOs to date reflects not any inherent safety in the technology itself but instead the success of a highly precautionary regulatory approach. Regulatory intervention is, indeed, one key to safety. In one famous case, when the Pioneer Hi-Bred company inserted a Brazil-nut gene into a soybean plant, hoping to improve the nutrient quality of soybeans for animals, the U.S. FDA worried about some of the beans leaking into the human food supply and advised conducting allergenicity tests on human subjects. These tests revealed that a Brazil-nut allergen had been transferred to the new GM soy, so Pioneer discontinued product development, and this potentially dangerous product never reached the market (Nordlee et al. 1996).

In this case, however, a risk was caught in time not by highly precautionary regulators in Europe but instead by regulators in the United States, where GMO technologies have always been regulated in much the same way that conventional food and crop technologies are regulated. In fact, most of the GMO technologies currently on the market have been put there by relatively lax American regulators rather than precautionary European regulators. Up to November 2006, regulators in the United States had approved seventy-seven different crops, compared to just twenty-seven in Europe, yet despite this larger number of American approvals, scientific evidence of new risks has not yet been found.

So long as genetic engineering is applied to the improvement of domesticated agricultural crop plants, a significant measure of safety to the larger biological environment is likely to be maintained. Even when they have been engineered to carry new genes, GM crop plants are still domesticated plant species that will not compete well in the wild. If left untended, farm fields are quickly smothered by invasive wild plant species with much greater competitive fitness—plants that have evolved to survive without the weeding, watering, and insect control provided by farmers. Whereas farm fields tend to fill up with weeds if humans do not intervene, untended wild fields and woodlands do not tend to fill up with domesticated farm crops, and GM varieties of corn, cotton, and soybean are still domesticated varieties. Even if genes from a GM farm plant were to outcross with a hardier wild relative of the plant, the progeny of the cross would almost certainly be less competitive in the wild because it would be inheriting a large set of thoroughly domesticated crop genes along with any advantageous transgenes that might come through. The most bioinvasive plant species are typically exotic wild nonagricultural plants (such as leafy spurge, introduced from abroad into grazing pastures in America).

Measuring the *inherent* safety of genetic engineering, or of any other crop science technique, is conceptually problematic. Conventional risk assessment considers hazards only on a case-by-case basis, and only when a technology is being used as its developer intended. For example, conventional risk assessment does not try to judge the inherent safety of electrical power; it only rates the safety of a specific electric power cable or battery when used in the intended manner. Electrical cables that are unsafe when plugged into the wrong power source can still be rated as safe in their intended use. Still, efforts have been made to rate the inherent safety of genetic engineering in farming. In 2004 the U.S. National Academy of Sciences (NAS) estimated the inherent potential for different genetic manipulation techniques to introduce unintended effects at the crop development stage. This study concluded that the crop improvement technique most likely to produce such unintended effects was not genetic engineering, but instead mutation breeding, a long-standing and noncontroversial technique. This report then went on to note that even conventionally bred crops can produce unintended and harmful effects, such as high levels of natural toxins. The NAS study pointed to past problems with various non-GM tomato, potato, and celery varieties that had to be withdrawn from the market because of elevated levels of natural toxins (NAS 2004).

Absence of Consumer Benefit

Citizens in rich countries dislike GM foods and crops not because these products carry new risks but instead because they so far have provided consumers with no new benefit. The first generation of genetically engineered crops released to the market in 1995–96 was designed to provide benefits primarily to crop producers rather than food consumers. The new genetic traits engineered into these crops allowed farmers to control insects and weeds with fewer and less-toxic chemical sprays and with less soil tillage, all of which reduced on-farm chemical input costs, fuel costs, and labor costs. The private biotechnology companies that developed this first generation of GM crops, led by Monsanto in St. Louis, targeted seed-buying commercial farmers as their first customers. They developed seed technologies with the needs of these farmers uppermost in mind. As for the food consumer at the end of the marketing chain, the companies assumed the absence of specific new consumer benefits would not be a problem. If the new GM seeds resulted in products that didn't cost any more, look any different, or taste any different; if they were not more difficult to prepare and delivered the same nutrient properties; and if they still measured up to the existing food and environmental safety standards, biotechnology companies (incorrectly) assumed that consumers would be willing to accept them.

When GM seeds first went on the market in 1996, seed-buying farmers immediately noticed and appreciated the production cost reductions and began using them as soon as they were made available. In the United States, the adoption rate for GM soybeans exceeded even the high rate of adoption half a century earlier for hybrid maize. GM cotton and GM maize spread nearly as fast. The gains farmers realized from these technologies were substantial. By one calculation, soybean farmers in the United States saved from $25 to $78 per hectare per year by switching to GM herbicide-tolerant soybeans, even after the higher costs of purchasing genetically engineered seeds were taken into account. Farmers who switched to GM insect-resistant corn reduced production costs by $5 to $9 per hectare, even with the higher seed costs factored in, and in addition they enjoyed an average yield increase of 5 percent. Farmers switching to GM insect-resistant cotton saw their profitability level increase by $53 to $116 per hectare. The cumulative net farm-income benefit in the United States between 1996 and 2004 was subsequently estimated at $10.7 billion (Brookes and Barfoot 2005). By 2006 farmers in the United States were using GM varieties on 89 percent of their acreage planted to soybeans, on 83 percent of their acreage planted to cotton, and on 61 percent of their acreage planted to corn (USDA 2006).

Farmers were not the only beneficiaries. Comparable commercial gains were made by the private companies that sold the GM seeds and by the innovators that licensed their patented technology. For GM cotton and soybeans, the gains captured by farmers represented 59 percent and 20 percent of the total economic surplus respectively, whereas the patent-holding innovator, Monsanto, took in 21 percent of the total surplus generated by GM cotton and 45 percent of the surplus generated by GM soybeans. Economic gains made downstream from farms by commodity purchasers in the food and fiber industry were substantially smaller. In the United States, downstream purchasers capture only 9 percent of the economic surplus generated by GM cotton and only about 10 percent of the surplus generated from GM soybeans (Falck-Zapeda, Traxler, and Nelson 2000).

This lack of significant downstream benefits made the technology an easier target. By the time these new products reached ordinary citizens at the end of the marketing chain, nearly all the economic surplus they had generated had been captured by someone else, and as for noneconomic consumer benefits, there weren't any. This made it easy for the critics of the technology to sway citizens in rich countries against it. In today's rich countries, consumer opinion is sovereign; the downstream consumer is always right. Farmers are not going to defend GMOs—or plant them—once processors, wholesalers, retailers, and consumers begin avoiding them. As for the seed companies, they were not going to defend something farmers wouldn't buy. This left (in addition to scientists) only the heavily invested technology innovators, such as Monsanto, vocally defending GM foods and crops. Monsanto in particular was poorly cast for this role; when it became the world's leading purveyor of GM crops in the 1990s, its global public reputation was still recovering from having earlier supplied American military forces in Vietnam with a dioxin-contaminated chemical herbicide, Agent Orange.

Survey research has confirmed the salience to consumers of benefits over risks. Only when the use of a genetic technology confers no added benefit will public attitudes tend to be strongly influenced by secondary concerns like views toward the company supplying the technology, or views toward the scientists who developed it, or views toward the government regulators who approved it (Hossain et al. 2003). Surveys also show that even in Europe, where sensitivity to perceived risk is high, consumers will still tend to allow the presence or absence of a benefit to dominate their final judgment (Gaskell et al. 2004). The easiest way to demonstrate the truth of this proposition is to compare GMOs in agriculture to GMOs in medicine.

Rich Countries Welcome GMOs in Medicine

Genetic engineering has been widely used in commercial medicine since 1982, when the FDA approved the first "bioengineered" drug, a recombinant form of human insulin. This drug was created by inserting the appropriate human gene into a bacterium, which then manufactured insulin as it grew, a vast improvement over the previous, more expensive method of deriving insulin from beef or pork pancreatic tissue. Then the first recombinant vaccine, approved in 1986, was produced by engineering a gene from the hepatitis B virus into yeast. In human medicine, valuable therapeutic proteins are now routinely manufactured from genetically engineered versions of living bacteria, yeast, or cultured mammalian cells. Genetically engineered cells from the ovaries of Chinese hamsters (known as CHO cells) are a favorite for recombinant drug production.

In the United States more than 130 recombinant drugs and vaccines have now been approved by the FDA, and in contrast to agricultural GMOs the pace of new approvals has been speeding up rather than slowing down, with more than 70 percent of all FDA approvals made in the past six years (Clearant 2006). In the European Union a total of 87 recombinant therapeutic proteins have been approved by the European Medicines Agency (EMEA), with 60 percent of all approvals made in the past six years (EMEA 2006). Approximately one-quarter of all new drugs coming on the market today are recombinants produced using GMOs, and the boom in GM medical drugs is likely to continue. In 2006 a New York–based market research company, Kalorama Information, projected that transgenically produced biopharmaceutical drug sales would increase by 140 percent in the next six years, to reach $12 billion by 2012 (*AgraFood Biotech* 2006).

This widespread use of GMOs to produce medical drugs generates little or no social or political controversy in rich countries. Surveys confirm that ordinary citizens support the use of modern biotechnology significantly more in medicine than in agriculture. According to one weighted survey study, in Europe modern biotechnology in medicine was supported by 59.3 percent of the population, versus a smaller 33.7 percent who support modern biotechnology in food. In the United States, according to this same study, 77.9 percent supported biotechnology in medicine compared to 57.7 percent who supported it in food (Priest, Bonfadelli, and Rusanen 2003). Broad citizen support for GMOs in medicine developed early in Europe and was little affected during the second half of the 1990s by the controversy over GMOs in food and agriculture. According to one survey of European citizens who had well-formed attitudes, support for GM foods fell from 61 percent to 47 percent between 1996 and 1999, but during the same period support for GM medicines remained strong, falling only slightly from 91 percent to 87 percent (Gaskell et al. 2000).

This divergence in social support for medical versus agricultural GMOs confirms that it is not the practice of genetic engineering that citizens in rich countries find objectionable but instead the purpose to which this science is being applied. When applied for the purpose of improving human health, it is strongly supported by citizens in rich countries, but when applied for the purpose of increasing the productivity of farming, it is supported far less. Such preferences make perfect utilitarian sense in rich countries where more agricultural productivity is scarcely needed, but constantly improving health outcomes and longevity are strongly desired.

The divergent social response to GMOs in medicine versus agriculture indicates other things as well. Just as we cannot blame this divergence on the use or nonuse of genetic engineering, neither can we blame it on public fears or misunderstandings of science, or on the role of multinational corporations in creating and delivering the products, or on the patent claims made by those companies, or on the high product costs that result from those patent claims, or even the suspect competence of the regulatory authorities that approve the products. All these factors are visibly present in the case of recombinant medical drugs, no less than in the case of genetically engineered foods and crops.

Public Fears of Poorly Understood Science?

The basic recombinant DNA (rDNA) techniques that lie at the foundation of both GM drugs and crops are equally mysterious to nearly all citizens in rich countries, few of whom have any training in modern molecular biology. Because the underlying science for both applications is the same and because public understanding in both cases is equally low, differences in technology acceptance cannot be explained using science comprehension as a variable.

Despite the greater salience of GM food issues in Europe, the science of genetic engineering is even less well understood there than in the United States. A Eurobarometer survey technique first developed in the European Union employs a set of eleven true or false statements to test public understanding of the science behind GMOs. The average score for Americans who took this survey was 64 percent correct, versus just 52 percent correct in the European Union. The questions are of the most elementary nature, yet only 57 percent of American citizens appeared to be aware that "ordinary tomatoes contain genes." In Europe an even lower 36 percent of respondents got this simple question right, and most Europeans believe erroneously that a person's own genes can be modified from eating GM food (Hallman et al. 2003, Gaskell, Allum, and Stares 2003). Yet it is not this slightly weaker comprehension of the science in Europe that explains Europe's weaker acceptance of GM foods and crops compared to GMOs in medicine, since the science of both is equally incomprehensible to consumers. Moreover, Claire Marris, a sociologist of science at the French National Institute for Agronomy Research, has been able to show that most ordinary citizens in Europe fully acknowledge their lack of comprehension in this area and do not use issues related to the science as the basis for their opinions one way or another (Marris 2001).

Mistrust of Globalization or of Multinational Corporations?

Explaining citizen reactions to GM crops as a possible backlash against corporate power and globalization is equally difficult, since the market for GM drugs is just as globalized and corporate-led as the market for GM crop seeds, and since social mistrust of multinational drug companies is just as widespread as social mistrust of biotech seed companies. Multinational drug companies can get away with being mistrusted because they deliver products with benefits widely valued in rich countries, whereas multinational seed companies do not. Large drug companies often make themselves hard to love. In 2001, in what the *Guardian* newspaper described as "one of the great corporate PR disasters of all time," a collection of 39 pharmaceutical companies filed a heavy-handed lawsuit against the Government of South Africa in an effort to protect patent claims on HIV/AIDS drugs. The resulting public outrage forced the companies to withdraw the suit. Drug companies have often been portrayed in popular culture before and since as iconic global villains, as in John Le Carre's 2000 novel *The Constant Gardener,* which is premised on the illegal actions of a greed-driven multinational drug company in Africa. Mistrust of the drug companies does not, however, spill over into a mistrust of the drug products, because the tangible benefits are valued by so many. In fact, the most frequent complaint against these companies is that they are not making their products available early enough or cheaply enough to a wide enough pool of beneficiaries.

High Product Cost?

The low popularity of GM crops in rich countries also cannot be blamed on the fact that these products are costly and protected by patents, since the same obviously goes for GM drugs. Cerezyme, an intravenous recombinant treatment (not a cure) for Gaucher disease developed by Genzyme, costs $200,000 *per year* for the average patient. Genzyme actually keeps a separate staff of thirty-four professionals working full time to help patients find insurance plans able to pay for the treatment. The total value of all prescription drug purchases in the United States tripled between 1980 and 2000 to reach approximately $200 billion per year, yet these costs are paid willingly by individuals or their insurance companies because of the valued benefits.

Mistrust of Government Regulators?

The greater social acceptance of medical versus agricultural GMOs in rich countries also cannot be attributed to factors such as mistrust of government in general or of government regulators in particular. Surveys do show that citizen fears about food safety will be highest in those countries where political institutions are trusted the least, yet this finding applies to fears about all foods, not specifically GM foods (Kjaernes, Dulsrud, and Poppe 2006). Moreover, this finding fails to explain the continuing support given to GM medicines even in low-trust countries. Survey research in Europe has yet to find any strong link between broad trust in government and a specific comfort or discomfort level with agricultural GMOs (Bocker and Nocella 2005).

Regarding the more narrow issue of untrustworthy governmental regulators, we have seen that mistrust in Europe of food-safety regulators following the 1996 BSE scandal spilled over to make citizens wary of newly approved GM crops. Yet this wariness never spilled over to affect citizen attitudes toward GM drugs. By 1999 only 47 percent of citizens in Europe supported biotechnology in food, but 87 percent still supported it in drugs (Gaskell et al. 2000). Drug regulators in Europe have had a dodgy track record for caution going all the way back to the hasty approvals given for the German morning-sickness drug thalidomide in the 1950s, a drug never approved for use in pregnancy in the United States and subsequently banned in Europe because of its teratogenic effects. Europeans are so forgiving of lax drug regulations that by 1999 they were citing the thalidomide case as a *positive* example of regulators being able to withdraw a product once it demonstrated a harmful effect (Marris 2001).

The current willingness of Europeans to trust the regulators of GM drugs more than the regulators of GM foods finds expression in an informal color code. Drugs made with rDNA have come to be referred to as "red" biotechnologies, distinguished from foods and crops developed with exactly the same science that have come to be called "green" biotechnologies (Bauer 2005). This arbitrary distinction has made it easier for Europe to impose stifling regulations on the GM crops that people don't like without having to block development of the GM drugs they do like (Falkner 2006).

In the United States GM medicines are also trusted more than GM crops, yet trust in regulators is not the reason, since both are regulated under the same relatively permissive "Coordinated Framework for the Regulation of Biotechnology," created in 1986 by President Ronald Reagan's Office of Science and Technology Policy (OSTP). Under the guidelines of this framework, new products developed using genetic engineering—including drugs, foods, pesticides, and crops—would not be treated differently, for regulatory purposes, from new products conventionally developed (Jasanoff 2005). In the United States, as a result, both medical and agricultural GMOs have been regulated under the same statutes earlier used to regulate conventional foods and drugs. Foods that are genetically modified do get closer regulatory scrutiny from the FDA than conventional foods, but through mechanisms that remain largely voluntary.

If regulatory stringency were the key to citizen trust, GM foods and crops would be trusted far more in Europe today than in the United States, but instead it is the other way around. From the start, Europe regulated agricultural GMOs with a separate set of more stringent statutes and procedures. The Single European Act of 1987 had launched a larger process of harmonizing regulatory standards around a high level of social protection, and as early as 1990 the European Union (then still the European Community) promulgated a separate directive (Council Directive 90/220/EEC) to regulate, according to a more precautionary standard, the deliberate release into the environment of genetically modified organisms. Although the various GMOs originally approved under this system in 1996 never led to any documented harms in Europe, standards were nonetheless tightened in 1997 through introduction of a mandatory labeling system for GMOs and then through a suspension of new approvals in 1998, pending a drafting of even tighter rules. These tighter approval rules were finally set in place in 2004, accompanied by an even more comprehensive labeling and tracing regulation.

Despite all this regulatory stringency in Europe, citizen support for GM foods went down rather than up. One opinion-survey analyst noted this anomaly in 2006, and offered an explanation: "Even the EU's recently overhauled regulatory framework for GMO authorisation and labeling has yet to make Europeans more accepting of food made from genetically engineered plants. In fact, only 27 percent of survey participants believe that the technology behind GM foods should be encouraged. It seems that most consumers have a hard time seeing any clear benefits associated with genetically engineered crops" (GMO Compass 2006, p. 1).

In Europe the politics of GM food acceptance is dominated in the end by benefit calculations, not risk calculations or trust calculations. Helge Torgersen at Austria's Institute of Technology Assessment uses comparisons to medicine to make this point: "Medical applications with substantial benefits get support despite some risk, but GM crops lack support even if risks are low" (Torgersen 2000).

To illustrate how much risk European regulators are ready to accept from GMOs in the medical area, consider the case of a new anti-clotting drug, ATryn, manufactured in the milk of genetically engineered goats—in fact, *American* GM goats. A U.S. company named GTC Biotherapeutics has developed this new drug technology using a herd of about 1,400 transgenic goats it keeps on a farm in central Massachusetts. A strict application of the precautionary principle would have dictated against approving a new drug extracted from the milk of gene-spliced goats, as this milk could be carrying potentially risky animal proteins, yet in June 2006 EMEA announced it would recommend approval in Europe of this new treatment. In this case the anticipated health benefits would not be especially widespread, since only one out of every three to five thousand people has the hereditary condition that leaves them without the needed anti-clotting protein in their own blood, but for those who did have this condition, the benefit would be substantial, so the precautionary principle was conveniently ignored. The only previous commercial source of the needed protein had been donated human blood, which was an expensive option. Thanks to milk production from the new GM goats, commercial prices would fall. A single one of these remarkable "pharm animals" would be able to produce about $150,000 worth of the protein every year (Heuser 2006).

Citizens also accept GM drugs more than GM foods and crops despite clear evidence, developed from clinical trials, that the drugs are more dangerous. Zevalin, a popular recombinant drug for non-Hodgkin lymphoma, can cause severely reduced white cell counts resulting in both gastrointestinal and respiratory complications; Avastin, used to slow tumor growth, can cause gastrointestinal perforations and hemorrhage; Raptiva, used to treat psoriasis, can leave the patient vulnerable to serious infection and malignancy. Yet all these drugs have been approved for sale both in Europe and the United States. The risky GM drug TYSABRI, developed in the United States for multiple sclerosis patients by Biogen Idec, was in such demand that it was given accelerated approval by the FDA in November 2004, only to be pulled from the market three months later following the occurrence in two patients of a brain infection known as progressive multifocal leukoencephalopathy (PML), in one case fatal. Yet due to the known benefits and popularity of this drug, it was brought back onto the market eighteen months later, this time under a slightly more rigorous risk-management plan.

Survey researchers like to divide those who support medical biotechnology into two different categories: those who support it because they see no risk, and those who support it despite the risks they see. In Europe in 1999, nearly half (47 percent) of all those supporting biotechnology in medicine said they were doing so despite the risk. There were actually more risk-tolerant supporters in Europe than in the United States, where only 35 percent of all supporters of medical biotechnology were positive despite seeing a risk (Gaskell et al. 1999).

Uncontrolled Exposure

In addition to the difference in anticipated benefit, there is another plausible explanation for why citizens in rich countries accept GM medicines much sooner than GM foods and crops. Risk theorists know that public acceptance is less likely when personal exposure to a risk is perceived as involuntary (Fife-Schaw and Rowe 2000). Exposure to GM medicines is something the individual can control, because the technology is physically contained inside laboratories, then carefully labeled and prescribed by physicians. In contrast, GM crops and foods are grown openly in the environment and then distributed through commercial channels, sometimes without labels, so that ordinary citizens might purchase them unknowingly.

Even a small or imagined risk can be felt as threatening when people are unable to control their exposure to the risk (Frewer et al. 2004). Lack of personal control over exposure helps explain the occasional public outrage when small amounts of fluoride are added to public drinking water supplies in an innocent effort by public health officials to prevent tooth decay. There still is no evidence of harm from fluoride, and ample evidence of benefit, yet because public exposure is essentially involuntary, many communities even today hold back from initiating the practice. Similarly, when GM foods first appeared on the market in Europe in 1996, consumers were outraged because there was not yet in place a system of mandatory labeling to give them a means to avoid consuming such foods. When exposure to a new technology is within an individual's control, as in the case of microwave ovens, cell phones, dangerous dietary supplements (like Ephedra), or GM medicines, fear and anger is usually neutralized because a personal choice to avoid the technology is always available.

Exposure control is important, but if control over individual exposure to GM foods were really the first concern, much greater anxieties about GMOs would be expected in the United States, where the technology remains unlabeled in the marketplace, rather than in Europe, where stringent labeling is now mandatory. Even though individual exposure can now be effectively controlled in Europe, consumers remain opposed to GM foods and still don't want them in the marketplace—because they have not yet seen any prospect of personal benefit.

When benefits are available, food consumers pay surprisingly little heed to uncontrolled risk exposures. Food can be dangerous to eat if carelessly handled or prepared, yet food consumers in rich countries are increasingly entrusting these vital tasks to complete strangers. In the United States alone, foodborne diseases routinely cause roughly 5,000 deaths every year, and they make an additional 76 million people sick (Mead et al. 1999). Yet Americans are preparing fewer and fewer of their own meals in pursuit of valued benefits such as greater convenience, food variety, and social pleasure. In Europe, a rapid increase in the number of women in the workforce has triggered a parallel rise in the consumption of processed and convenience foods, all prepared by others. In France time spent on meal preparation at home has fallen by half since the 1960s, and fast-food restaurants are on the rise even in Greece, Portugal, and Sweden (Mitchell 2004).

American and European food consumers today value access to variety and convenience much more than personal control over handling and preparation. American supermarkets carry as many as 400 different produce items, up from an average of just 150 different items in the 1970s (Putnam, Allshouse, and Kantor 2002). Hypermarkets like Wal-

Mart offer one-stop discount shopping with plenty of parking space, but little or no consumer control over the provenance of the food items being purchased. Consumers also relinquish control over food preparation when they purchase heat-and-serve foods and when they take their meals in restaurants. For two-income households, single-parent households, and working individuals who live alone, the convenience of heating and serving foods prepared by others is viewed not as an uncontrolled risk exposure but as an upscale amenity. Restaurant patrons have no way of knowing how long the canned cheese sauce has been out of the refrigerator, or if the workers in the kitchen have failed to separate the meat from the vegetables on the cutting board, forgotten to wash their hands, or perhaps come to work sick and sneezing. Restaurant kitchens are seldom inspected by health officials, yet they are almost never visited by patrons. American restaurant-goers are fully aware that official safety inspection systems are weak: 27 percent actually *believe* restaurants to be the primary source of foodborne illness, yet the pursuit of a tangible direct benefit—the undeniable convenience, variety, and social pleasure they take from enjoying a meal at a restaurant—is far more important than even a known food-safety risk. In the United States, by 1997, roughly 40 percent of all food expenditures went for meals prepared by others rather than at home. In the United Kingdom the share was 27 percent and in Spain, 26 percent.

So in the end it is the absence of a compelling consumer benefit that has done most to undercut social support for GM foods and crops in rich countries. Margaret Mellon, director of the agricultural and biotechnology program of the Union of Concerned Scientists in Washington, D.C., and a prominent skeptic regarding GM crops, admits candidly that the meager consumer benefits are more important in rich countries like the United States than any hypothetical risks:

> [M]y colleagues and I at the Union of Concerned Scientists are not opposed to biotechnology. We think its use in drug manufacture, for example, makes a lot of sense. The therapeutic benefits of the new drugs outweigh the risks, and often there aren't any alternatives . . . Agricul-

ture isn't like medicine. We in the U.S. produce far more food than we need. And we are so wealthy that whatever we can't produce we can buy from somebody else. As a result there are 300,000 food products on our grocery shelves and 10,000 new ones added every year. The notion that consumers in the U.S. fundamentally need new biotechnology foods isn't persuasive. (Mellon 2001, p. 64)

What about GM Foods with Consumer Benefits?

The first generation of GM agricultural crops was designed with improved agronomic traits valuable to farmers, but with nothing new for consumers. A second generation of GM crops is just now emerging from the research pipeline, designed to give specific food benefits to consumers, such as enhanced beta-carotene content (for vitamin A) or higher omega-3 content (found in heart-healthy seafood diets). The private biotechnology industry hopes the delivery of these clear benefits to consumers will break down the high social resistance GM foods and crops have met so far (Shoemaker, Johnson, and Golan 2003), but such prospects are far from certain. Nutrient traits such as beta-carotene could be quite a valuable benefit if engineered into the nutrient-deficient staple food crops that dominate the inadequate diets of poor people in the developing world, but they are unlikely to be highly valued by consumers in rich countries, who have plenty of other ways to get enough vitamin A or omega-3 oils.

Consumers in rich countries today do not need GM foods to attain adequate nourishment, since they already have a diverse diet of inexpensive and nutritious non-GM foods available, as Margaret Mellon pointed out. American diets are notoriously deficient in adequate daily servings of vegetables and fruits, yet this is not for lack of access—remember the 400 different produce items that are currently available in each American supermarket. Any American or European citizen who cares to be serious about dietary improvement can achieve this goal without having to consume any GMOs, and surveys confirm this is how most would prefer to proceed—eating conventional foods, not GMOs. Attitudes on this subject are particularly strong in Europe; when asked if they would continue buying a nutritionally enhanced food if they learned it was genetically modified, a majority of survey respondents in Germany, Australia, and Great Britain said they would stop buying the food (Hoban 2004).

GM Foods and Crops in a Larger Context

In addition to the lack of consumer benefit and anxieties regarding uncontrolled exposure, citizens in rich countries today tend to disapprove of GM foods and crops because of what they think this new technology will do to farming. In rich countries today, farmers are already highly productive, perhaps too productive, as Margaret Mellon suggests. New applications of science that could bring still more productivity have become suspect and stigmatized as a result. Long before GM foods and crops came along, citizens in rich countries had already become concerned about modern, high-yielding crops grown on specialized "factory farms" with heavy applications of chemical fertilizers or pesticides. Citizens had already come to believe that too much modern agricultural science was moving society away from traditional, small-scale family farms, which they perceived as socially valuable and working in harmony with nature. They did not want the growing of food to become just another industrial enterprise. In the chapter that follows I explore this turn against agricultural science in greater detail. I find that well before the era of GMOs, growing numbers of citizens both in Europe and the United States had come to believe farming was productive enough and government should pull back from using taxpayer money to promote still more agricultural science. This was a priority change completely understandable and affordable in the United States and Europe, but dangerously inappropriate if exported to Africa.

2

Downgrading Agricultural Science in Rich Countries

Genetically engineered crops are only the latest in a progression of science-based innovations that have brought stunning changes and unprecedented wealth to farming in the industrial world. In both the United States and Europe during the middle years of the twentieth century, powerful new applications of mechanical, biological, chemical, and information science were taken up by farmers, completely transforming the industry in little more than the length of a human lifetime. Fields once plowed and tilled by farmers walking behind simple teams of horses came to be groomed by giant 400-horsepower machines guided by satellites from space. Thanks to improved crop biology and fertilizers, cornfields that once yielded 30 bushels per acre now yield five times that amount. Chickens that were once sold scrawny after a comfortable life pecking for grubs in the barnyard are now fattened with carefully prepared rations as quickly as possible, and in factory-like confinement. Millionaire farmers with college degrees now oversee these highly specialized, industrial-scale production systems, spending more time on the computer than in the field or on their tractor.

The economic advantage of a highly productive, technologically advanced farming system is its capacity to bring down the cost of producing food while simultaneously releasing workers for employment outside of agriculture in the industrial and service sectors of the economy. The postindustrial affluence enjoyed today by urban and suburban dwellers in both America and Europe rests on those regions' much earlier adoption of science-based farming. And agricultural science has shown that it can continue to deliver new productivity gains, as in the case of genetically engineered farm crops. Yet the social response to science-driven farm productivity growth in rich countries today has turned from sweet to sour. In both Europe and the United States, many fewer citizens now benefit directly from added farm productivity gains, because many fewer citizens are now farmers. The slightly lower commodity prices that result from added farm productivity gains are of little interest to consumers because crop prices have long since dropped to bargain-basement levels, especially when compared to much higher average personal incomes in rich countries today. Urban consumers in rich countries fret about eating too much food and becoming obese rather than about having too little and going hungry.

As might be expected under such circumstances, consumers in rich societies begin demanding other things from the farm sector. They demand social values—such as foods grown locally by small family farmers rather than by industrial-scale farmers half a continent away. Or they demand health and environmental values—such as foods grown without the use of certain chemicals. Or they demand animal welfare values—such as pigs and chickens raised with access to outdoor spaces rather than in cramped confinement. New applications of agricultural science cease to be attractive, as they only seem harbingers of a still more heavily engineered and corporatized approach to crop and animal production. Quality-conscious citizens in rich countries at this point begin to desire less modern science and technology in their food production systems rather than more. In both Europe and the United States today, urbanites and suburbanites still admire farmers (as seen in their willingness to tolerate lavish subsidies for the sector), yet they have little interest in spending tax money to boost the productivity of farms. Public investments in agricultural science in rich countries have consequently stopped increasing, and in some instances they have even begun to decline.

In this chapter I trace the origins and dimensions of this recent change in public policy priorities in Europe and North America. I document the contributions of agricultural science to farm productivity and social prosperity in both Europe and North America during the twentieth century, and I mark the massive labor migration out of farming that accompanied and facilitated these changes. It was natural to expect, as this transition was being completed during the second half of the twentieth century, that several forms of social resistance would emerge against moving the process faster or still further, and that the importance attached to new investments in agricultural science would consequently be downgraded. I note this did indeed happen in both the United States and Europe, beginning in the 1980s. This downgrade in the priority given to agricultural science made sense in rich countries, where accomplished technology gains had already made the sector highly productive. Such a downgrade would not have been appropriate in these same countries earlier in the twentieth century, just as it is not appropriate today for science-starved Africa.

Growth in Farm Productivity in the United States and Europe

Both Europe and North America saw their farm sectors transformed in the twentieth century by new applications of science, including the new science of powered machine technology, chemical science, biological science, and information science. Because farming always combines biological with mechanical processes, it actually presents more numerous options for applying new science compared to the manufacturing or service sectors. Strong productivity growth in farming in rich countries throughout the twentieth century was one result. Sometimes farming is referred to as a sector that is "declining" or even "failing" in rich countries because it employs fewer people, but in terms of productivity—the value of output relative to the value of input—the sector has become sensationally successful in the last century and has been able to provide a massive benefit to the rest of society as a result.

Reliable measures of productivity trends in agriculture are sometimes difficult for economists to construct because of data scarcity, differences in product quality over time, possible confusion of data because some outputs like animal feed crops also serve as on-farm inputs, and the difficulty of accurately accounting for labor in a cottage industry in which multiple household members, including children, contribute their labor. Fortunately in the United States, consistent and meticulous data collection by various agencies of the federal government since the early years of the twentieth century make possible reasonable productivity growth estimates. Bruce Gardner, a leading American agricultural economist, has assembled agricultural input and output data for the years since 1910 and has presented them in a dramatic index form, as shown here. The accompanying figure shows that while the yearly value of total inputs in American agriculture (discounted for inflation) has remained essentially unchanged over the past eighty years, the yearly value of total outputs has increased by roughly 300 percent.

What made this spectacular productivity growth possible in the twentieth century was the uptake on farms of an impressive stream of scientific and technological innovations. Between 1900 and 1940 American agriculture began to use power machinery (tractors, drainage pumps, electric poultry equipment), new chemical applications (chemical fertilizers), new land-use systems (terracing and contour plowing), and new applications of biological science for both crop and animal production (hybrid corn, artificial insemination). During the second half of the twentieth century the pace of on-farm technology uptake accelerated still more, first with a nearly universal electrification of farming, then the development and use of new chemical treatments to control weeds and insects, then applications of information and computer science that brought improved management and marketing efficiencies to farms, and then new sensor technologies such as lasers, for the precise leveling of fields, and GPS technologies with satellite tracking and onboard computer monitoring to assist in more precise chemical applications. All these powerful changes were in place before the mid-1990s, when the first genetically engineered crops were finally introduced for commercial use.

The productivity of European agriculture has always tended to lag a bit behind the United States as a result of America's much greater endowment of naturally productive land and water, yet growth rates in science-driven agricultural productivity in Europe have been nearly as high as in America, and for the same reasons. A slightly different technology path was followed in Europe, where labor is more abundant than land. This circumstance favors innovations that enhance land productivity sooner than labor productivity, meaning that in Europe applications of chemical and biological science, which boost crop and animal

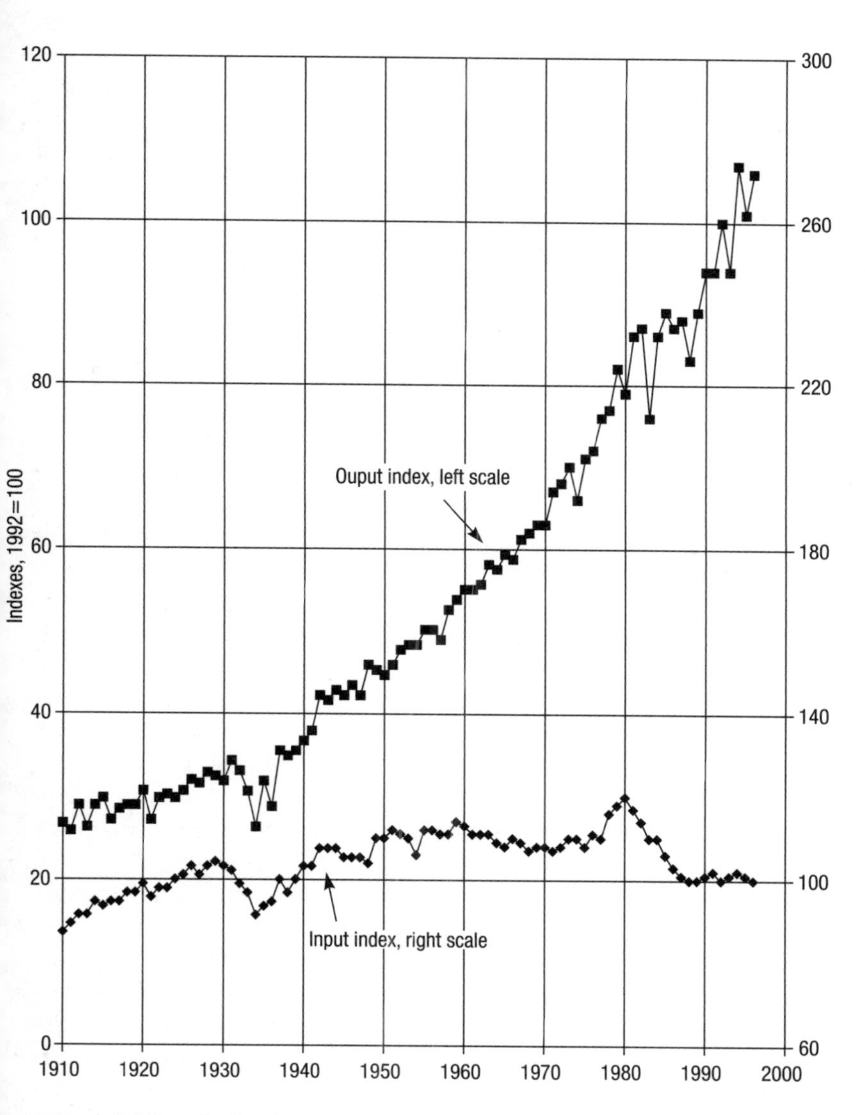

Indexes of U.S. agricultural output and inputs. Data from U.S. Dept. of Commerce and Council of Economic Advisers. Reprinted with permission from Gardner (2002).

yields, were originally favored over high rates of mechanization, which mostly just replace labor. Yet the productivity of both farm land and farm labor have seen significant increases on both sides of the Atlantic. Over the ninety-year period between 1880 and 1970, agricultural output per worker in the United States increased at an average annual rate of 2.81 percent, only slightly higher than the 2.35 percent rate in France, 2.37 percent in Germany, and 2.46 percent in Denmark. Output per hectare of land in the United States grew at an annual rate of 0.72 percent, only slightly lower than the 1.4 percent rate in France, 1.64 percent in Germany, and 1.67 percent in Denmark (Ruttan et al. 1978).

Because demand for farm products does not grow as rapidly as demand for other products and services in affluent societies, increased productivity on the farm eventually allows a reallocation of assets to other sectors. Valuable labor and investment resources begin shifting out of farming. In terms of its share of the labor market, in particular, farming declines dramatically. In the United States, the percentage of citizens employed in agriculture fell from more than 50 percent in 1870 to only 2.8 percent by 1990. In France during the same period, the percentage of farmers fell from 50 percent of the workforce to 6.1 percent, in Germany from 47 percent to 3.4 percent, and in Denmark from 48 percent to 5.6 percent. This dramatic movement of farm labor into the urban economy facilitated further growth in the manufacturing sector, where parallel applications of new science were also at work boosting productivity and income.

In rich countries in the twentieth century most economic growth across the board has been driven by scientific advances. Charles I. Jones, an economist at the University of California at Berkeley, has calculated that between 1950 and 1993 roughly 80 percent of all economic growth in the United States was attributable either to "increased research intensity" (50 percent) or to rising levels of educational attainment (30 percent) (Jones 2002). How much productivity growth in the farm sector, specifically, was attributable to science? Clearly other factors were also important, including higher educational attainment and public investments in rural infrastructure (e.g., power and roads), but agricultural economists have concluded that the single most important factor, by far, was technological change. Wallace Huffman and Robert Evenson (1993) found that 95 percent of all farm productivity growth between 1950 and 1982 was attributable solely to the lagged effects of scientific research, plus extension of the results of that research to farmers. Munisamy Gopinath and Terry Roe (1997) examined data between 1949 and 1991 which showed that spending on agricultural research was the single most important contributing factor to productivity growth.

Impacts in the United States

Over the course of the twentieth century in America, the total area of harvested cropland changed very little, yet the average size of individual farms more than tripled. Small, horse-powered farms were progressively consolidated into larger, fully mechanized farms, and by the end of the century 58 percent of all U.S. agricultural sales were being made by large commercial farms that averaged 1,834 acres in size.

Farm acreage continued to dominate most country landscapes, yet in fact rural America was becoming far less agricultural during the twentieth century. Actual rural depopulation was rare, since the massive departure of labor from farming was more than offset by a diverse influx of nonfarmers, seeking the obvious amenities of country life. Nonfarming, country-dwelling Americans increased roughly fourfold in the twentieth century—and they now actually outnumber farmers in rural America by more than ten to one. In fact, farm residents are almost as small a minority in rural America today (7 percent) as they are in the nation as a whole (2 percent). As recently as the 1930s more than three-quarters of all rural counties in America depended on agriculture as their primary source of income, but today only 20 percent of rural counties depend on agriculture for 15 percent or more of earnings (Offutt and Gundersen 2005). Small towns in rural America are much less agricultural today, but they remain as numerous as ever. The total number of rural dwellers living in small towns of less than 2,500 in America has actually increased slightly over the course of the past century. It is true that in some regions small agricultural towns have been boarded up, especially on the Great Plains, but these losses can just as well be traced to nonagricultural changes such as the introduction of interstate highways, which encouraged rural commuting to more distant central cities (Gardner 2002).

While the farm population in America was losing numbers due to productivity growth, it was making striking gains both in income and accumulated wealth. In 1920 the average household income of farmers was only 40 percent as high as the average for nonfarmers; by the 1990s average farm-household income in America was equal to or exceeded that of nonfarm households. Meanwhile farm real estate in America—land and buildings—increased roughly 50 percent in value per acre over the course of the century. Largely due to the increased value of farm assets (worth more because they had become more productive), the average net worth of American farm households in 2003 was $664,000, compared to just a $90,000 average for all households in America (Offutt and Gundersen 2005). As recently as 1959 more than half of all people living on farms in America fell below the official poverty line, but by 2000 only 10 percent of Americans living on farms fell below the line, which was a lower poverty rate than for nonfarm dwellers (ERS 2005). Those Americans who remained in farming were now better educated as well. Between 1950 and 1980 the proportion of farm-dwelling American males holding a high-school degree quadrupled, almost catching up with the growing proportion of urban-dwelling graduates.

Thanks to science-based productivity growth, these dramatic gains in wealth and welfare on the farm were made without anyone having to pay higher prices for the goods produced on farms. In fact, between 1900 and 2000 the crop and livestock prices received by farmers in America declined in real terms by more than half. Farmers were able to prosper despite this decline in crop prices because their production costs declined at an even faster rate as a result of their use of the new technologies. In the years after 1940, multifactor productivity in American agriculture was increasing at an annual rate of about 2 percent, which helped to bring an annual 1.8 percent decline in production costs; this steep cost decline was more than enough for farmers to remain profitable despite the annual 1.5 percent decline in prices received (Gardner 2002).

There is no such thing as an average farm in America today. Outdated census rules continue to count as a farm any place that produces and sells as little as $1,000 worth of agricultural goods each year, a lax standard that allows many semi-retired rural dwellers to be counted as farmers even if they only keep a few dozen sheep as a hobby, sell a horse, or maintain a small orchard for a roadside stand. The 2 percent of all Americans who count as farmers by this lenient definition differ as much from each other as they do from nonfarmers, yet from the perspective of their adoption of new technologies they can be lumped into three broad categories: full-time farmers who lead in technology adoption and are doing extremely well; full-time farmers who lag in technology adoption and therefore are not doing as well; and part-time farmers who care little about new farm technologies because their household income does not depend on their productivity as farmers, or their sale of farm products. This third group is now numerically the largest by far, and the second group attracts the greatest social sympathy, yet it is the first group that has come to supply most of the nation's agricultural production.

Roughly three-quarters of all American farms are classified as small, selling less than $50,000 worth of products every year. For most of these small farms, agricultural production is a secondary occupation, often just a hobby, and seldom profitable on its own terms. In 1995, the average net cash income these households received from farming was actually *negative* $2,900 (Orden, Paarlberg, and Roe 1999). What keeps these so-called farms financially viable is the average $30,000 in family income earned off the farm. In aggregate commercial terms these small farms are barely significant. They provide only about 10 percent of the nation's total agricultural output by value.

At the other end of the diverse spectrum of American farms we find the 6 percent of all farms falling into the $250,000 (or above) annual sales class. This small fraction of total farms contributes 58 percent of all annual farm-product sales in America, and they average more than $300,000 yearly in net cash income from farming. The growing commercial dominance of these larger and more specialized farms shows up clearly in measures of market concentration. Between 1930 and 1992 the disparity between the market sales of the largest 10 percent of

American farms versus the smallest 50 percent increased twentyfold. The largest 10 percent now sell 190 times as much as the smallest 50 percent (Gardner 2002).

A distinguishing feature of these commercially dominant large farms is their early adoption of new, productivity-enhancing technologies. Some new chemical and biological technologies in farming are essentially scale-neutral and commercially attractive for small as well as large farms, but purchase of the most powerful new mechanical technologies, or the most expensive new computer and remote-sensing technologies, usually pays off only for the largest and most heavily capitalized farms, a fact that drives American farms not only toward larger size but also toward greater specialization (Ruttan 2001). Highly diversified farms in America—those growing a wide variety of crops and raising several different kinds of livestock—have progressively been replaced by farms growing only one or two crops over a much larger area of land, or raising only one kind of animal in much larger numbers.

Parallels in Europe

Large, highly specialized farms developed in a similar manner in Europe over the course of the twentieth century, albeit from a different starting point. Applications of modern agricultural science have allowed European labor to move from farming to other sectors, farm size to increase, farm income to rise, and food prices to fall.

Europe began the industrial age with a relatively dense and remarkably impoverished farming population, a disadvantaged peasantry working the land on fragmentary holdings that often had changed little since feudal times. Whereas Americans could look West to a vast frontier of unplowed virgin land—including some of the best land for field crops in the world—Europeans faced a relative shortage of good farming land, especially in the drier southern regions. For this reason they concentrated more on applications of chemical and biological science, which boost land productivity, and less on mechanization, which replaces labor. Europe had the best scientific researchers in the world in the nineteenth century, and it was a German scientist, Justus von Liebig, who first mapped out the chemistry of soil fertility, a breakthrough that led to productivity gains through increased applications of naturally available phosphate, potash, and then guano and nitrate fertilizers imported from Latin America. Important innovations were also made in animal feed technologies in England, Denmark, and the Netherlands, which allowed reduced grazing and freed up vast areas of land for the production of higher-value crops. Crop rotations were scientifically improved on big Prussian estates, assisted by modern drainage systems copied from England (Tracy 1989). Beginning with a research station in Saxony in 1852, governments in Europe began making conscious investments of state funds in agricultural science. Between 1870 and 1914 most other European states adopted this "German model" of promoting farm productivity gains via public-sector science investments (van Zanden 1991).

European agriculture, with its more limited land endowment, was less competitive in a free market than American agriculture, a fact that first became obvious during the second half of the nineteenth century when sharply reduced global transportation costs—thanks to steam-powered ships and railroads—opened European markets to imports of cheap grain from the United States. These cheaper imports plus deflationary macroeconomic conditions threw much of European agriculture into a prolonged economic crisis, severely compounded during the first half of the twentieth century by two catastrophic world wars waged on European soil. It was not until the second half of the twentieth century that farming in Europe began a strong comeback, with a flood of new agricultural science applications playing a lead role.

Following the Second World War, industrial recovery in Europe made widespread farm mechanization a viable option, and the total number of powered tractors in Western Europe rose from half a million to 3 million between 1947 and 1960 (Tracy 1989). Nonproductive farm workers left the land to find industrial jobs in town, and as a consequence, farm output per worker in Europe actually increased more rapidly than industrial output between 1949 and 1959. Output per man in agriculture increased by 77 percent in Denmark, 72 percent in Germany, and 69 percent in Italy. Productivity growth in European industry later outpaced agriculture between 1960 and 1970, but farm production continued to surge ahead as still more scientific innovations were taken up in the agricultural sector. Increased fertilizer use during the decade of the 1960s allowed agricultural output per hectare to increase by 35 percent in Germany and the United Kingdom, and by 49 percent in France. Thanks to increased productivity, the income of farmers went up as well; between 1964 and 1972 real income per labor unit in agriculture grew by 9 percent in Luxembourg, 24 percent in the Netherlands, 28 percent in Germany, 59 percent in Belgium, and 75 percent in France.

Paralleling the U.S. model, Europe's increasingly productive farming economy also became more integrated with the nonfarm economy, which gave rural dwellers in Europe new options for high-paying work off the farm. The arrival of more weekend tourists and builders of second homes brought money from the prosperous urban economy out into the countryside. Farm employment in Western Europe fell from 30 percent of total employment in 1950 to just 4.9 percent by 2004, but total farm output continued to grow over the second half of the twentieth century at an annual rate of roughly 2.2 percent, reaching an aggregate value of €300 billion by 2004 (Josling and Tangermann 2006).

Little Resistance to Agricultural Growth from Farmers

The new wealth brought to American and European farming by science was welcomed, but the accompanying cultural and demographic changes were socially difficult. The movement of labor out of farming required a sometimes painful identity change for those who left their familiar agrarian culture to seek work in town. Farmers called it being "tractored off." Agrarian traditions suffered further insult when the surviving farms in the countryside abandoned the diversified cottage-industry model and became highly specialized, industrial-scale "factory farms." The much higher productivity of these specialized operations was welcomed, but only up to a point. Once the goods produced by these farms became so abundant and cheap as to be of little additional value to nonfarmers—who had an appetite for only so much food—investments in still more science to push the productivity of farming still higher began to encounter social resistance. Yet surprisingly, this resistance almost never came from farmers themselves.

Most commercial farmers welcome productive new technologies, since they—as owners of the land being made more productive—will be the first to profit. Even if they are not the first to take up a new technology, market pressures will soon induce them to do so once their neighbors start taking it up. If the new technology is based on biological or chemical science (for example, new hybrid seeds or nitrogen fertilizers), it will be affordable for use on small as well as large farms, so nearly all are likely to become eager adopters. High-yielding hybrid corn was first offered to American farmers in the mid-1930s, and within a decade it was being planted on small as well as large farms, in fact on 90 percent of total corn acreage (Fitzgerald 1993). Likewise, inorganic nitrogen fertilizers reduced the cost of soil nutrient replacement on both small and large farms, so both adopted the use of this new technology quickly. Social resistance to intensive chemical use on farms today comes primarily from nonfarmers.

The only application of modern biotechnology to have stirred any significant resistance from farmers in America so far has been a genetically engineered hormone (recombinant bovine somatotropin, or rbST) introduced in the 1980s to enhance milk production per cow. In the 1990s some smaller dairy farmers in the upper Midwestern states joined in a brief and unsuccessful effort to block this new technology. They feared it was not scale-neutral because it was being taken up more quickly by large industrial dairy operations in California. Yet the key leaders in this political initiative were nonfarmers: consumer advocates worried about the safety of the milk supply and animal-welfare advocates worried about an increased incidence of udder infections in cows (Ruttan 2001).

Power-driven mechanical innovations should more likely be resisted by farmers, because of the difficulty smaller farms have in taking them up, yet even these were welcomed rather than resisted by most smaller family farms during the twentieth century because they permitted reductions in the cost of hiring seasonal labor. The hired farm workers threatened by these technologies might have wanted to resist them but they were seldom in a social or political position to do so. When cotton harvesting was mechanized in the American South after World War II, millions of traditional hand-pickers—mainly poor African Americans—were displaced. In this case the technology was not resisted because the displaced African-American workers were still socially marginalized under a Jim Crow legal system and silenced through terror and intimidation. Resistance also failed to develop because the workers had an exit option. Just as the mechanical pickers were being introduced into the South, factory jobs were opening up in the booming postwar economy in the North. The resulting northward labor migration has been called a "second great emancipation" for African Americans, although some insist these workers were pushed off the cotton farms by mechanization at least as much as they were pulled away by new employment opportunities in the nonfarm sector (Holley 2000; Grove and Heinicke 2003).

Hired workers were also displaced by the introduction of mechanical tomato harvesters in California in the 1960s, an innovation that reduced the number of man-hours required per acre to harvest tomatoes by 56 percent. Yet there was again little resistance in this case because the laborers displaced had second-class status as ethnic minorities or seasonal, nonunion migrants. In strong contrast, unionized dockworkers were able to mount strong resistance to mechanization on the American waterfront in the 1950s.

Farm owners, for their part, almost never mobilize to resist productive new technologies. On one obscure occasion in 1869 a newly introduced wheat combine was burned in California, but the act was promptly denounced by a local newspaper, which called on all good men in the community to "unite and hunt up the offenders and make them feel the heaviest penalties of the law for damages and then be driven from every civilized community" (Schmitz and Seckler 1970, p. 569). Some farmers during the first half of the twentieth century did keep their horses rather than switching to motorized tractors, but more because they could not afford this attractive new innovation rather than as an act of resistance. The emergence in the 1990s of automated, industrial-scale concentrated animal feeding operations (CAFOs) presented a threat to the livelihood of traditional hog farmers in Iowa, but rather than trying to block CAFOs, they formed new cooperatives to offer their "free-range" pork to niche markets at a higher price (Grey 2000). Concentrated animal feeding operations face significant opposition in America, and for good reason, but almost all the resistance comes from nonfarming neighbors, or defenders of animal rights and the environment.

Farmers in America have the means to resist trends they do not like, as they are politically organized and far from passive. In fact, farm organizations in the United States have a proud and colorful history that includes resistance to almost everything except productive new technologies. They have fought against railroads, banks, grain companies, meatpackers, consumer advocates, big-city politicians, labor unions, animal rights advocates, and of course environmentalists (Rasmussen and Stone 1982). Nor do they have much use for East Coast college professors. Yet commercial farmers take a different view when it comes to their technology providers, including the big companies that supply them with seed, chemicals, and machinery. The free ball caps handed out by these companies are worn with pride.

In Europe as well, farmers have rarely resisted the new seeds, machines, and chemicals brought to them by science throughout the twentieth century. Once again it is not for any lack of capacity to resist. Farmers in Europe during the twentieth century were even better organized than their American cousins to enter the political arena and take action against whatever it was they did not like, often through open civil disobedience, leading at times to vandalism and violence. European farm activists routinely hold politicians hostage by driving their tractors or herding their animals onto highways and urban streets, or even airport runways. Their goal, however, is usually greater state protection against market forces; they have no particular history of taking political actions against new technologies. This has been the case in France since the days of Jules Meline, the great nineteenth-century champion of agricultural fundamentalism who once stated explicitly, "Our agriculture must improve its methods, perfect its techniques, and become scientific" (quoted in Tracy 1989, p. 73). European farmers fear overproduction and complain about falling prices, but one remedy they usually demand is a greater state effort to promote dissemination of the most modern equipment and latest techniques.

In the highly publicized case of GM crops in France, the activist groups that on occasion invade farm fields to uproot and burn transgenic crops are not usually led by farmers. Antiglobalization activist José Bové, who presents himself as a simple French sheepherder, is more accurately described as a professional social activist. Bové spent the first seven years of his childhood in Berkeley, California; the *Wall Street Journal* once labeled him a Bakunin-quoting former hippie who only became a sheep farmer in 1975 as a political act. He has such a busy international travel schedule and spends so much time either in court or in jail that his sheep do not see him for weeks at a time (BBC 2002). Although Bové enjoys great popularity among the nonfarming French, authentic commercial farming organizations condemn his direct actions against new technologies. In September 2006, after some 200 activists destroyed six hectares of commercially grown GM maize on a farm near Toulouse, roughly 800 farmers from the region marched through a nearby town to demonstrate against the attacks and present a petition demanding more government support for research and new technologies. In August 2007, when Bové led protesters into a farm field near Verdun-sur-Garonne, France's largest farm federation, FNSEA, mobilized an impassioned counterdemonstration against Bové that had to be restrained by gendarmes and teargas. Michel Masson, the regional FNSEA leader, warned Bové that real French farmers were nearly ready to "take their rifles off the wall" (Lichfield 2007).

Both in Europe and North America, legitimate farm organizations know better than to waste their time and energy resisting new technologies, recognizing this as a losing strategy going all the way back to the original Luddites. Farmers choose instead to use their substantial political clout to pursue more attainable goals from the state, such as tax breaks, relaxed environmental regulations, subsidy payments, and border protection schemes. Subsidy and protection demands from farm organizations are a universal in the developed world; they always emerge after industrial growth takes off, once farmers begin to sense they might be losing some comparative advantage to industry in the economic marketplace. Their response is always to organize and make demands in the political marketplace for an array of income-support devices, almost always with success (Honma and Hayami 1986). The average annual value of all income supports provided to farmers by governments in wealthy countries reached a remarkable $273 billion in 2003–05.

These government supports are now, in fact, the source of roughly 30 percent of all farm income in the industrial world (Anderson 2006).

Resistance to Science-Based Farming from Nonfarmers

In rich countries today resistance to science-based farming comes almost entirely from nonfarmers. New investments in agricultural science are of little direct value to nonfarmers because productivity on farms is already high enough to ensure the price they pay for food will be low relative to income. Thanks to science-driven farm productivity growth in America over the course of the twentieth century, the real cost of farm commodities fell by roughly 50 percent. At the same time, average consumer income increased in real terms by more than 400 percent (Gardner 2002). This drop in prices and growth in income has left the average (nonfarming) consumer mostly indifferent to any promise of still-lower farm commodity prices that more science-driven productivity growth might bring.

The diminished importance to consumers of added farm productivity is most commonly measured in terms of food costs as a percentage of total consumer income. In 1901 the average American consumer spent 41 percent of personal income on food, whereas today the average American consumer spends only 12 percent of personal income on food. This measure is dramatic enough, but it considerably understates the lost salience of farm productivity because it fails to incorporate the equally diminished role that farm commodity costs now play in final food prices paid by consumers. Before World War I in America, the products sold by farmers made up roughly 45 percent of the total cost of food to the final consumer (counting store and restaurant sales together), the other 55 percent was accounted for by storage, transport, advertising, processing, packaging, labor, and shelf space. Today, only 25 percent of the final cost of food consists of products sold by farmers, reflecting both much lower commodity prices and far more elaborate investments in processing, packaging, and advertising (Gardner 2002).

Combining these factors, we can calculate that a century ago 41 percent of income went to food, and 45 percent of that food spending went to pay for farm commodities, so the final share of personal income that went to pay for actual farm commodities was 18 percent. Today, by contrast, since only 12 percent of income goes to food and only 25 percent of that goes for farm commodities, just 3 percent of personal income goes for actual farm commodities. By implication, the average consumer today is only one-sixth as sensitive to the lower commodity prices that accompany farm productivity gains as a century ago. Correspondingly, consumers in rich countries today are also less sensitive to higher commodity prices. When a mandated expansion of corn use for ethanol doubled corn prices in North America in 2007, poor consumers in Mexico City staged a mass protest, but American food buyers barely noticed.

Prior to this science-driven drop in the cost of farm commodities, hunger in America had been a real problem, especially among the rural poor. Today, in part because food is so cheap, concern over hunger has been replaced by concern over an epidemic of obesity. Only one-half of one percent of American households now experience, on any given day, what the government currently measures as "hunger"—the inability of one or more household members to afford enough food (USDA 2004b). Rather than suffering from hunger, three in five Americans today are overweight. The most important causes of obesity include sedentary lifestyles, reduced food preparation time thanks to microwave ovens, unhealthy convenience foods, more meals taken at fast-food restaurants, and much less cigarette smoking, yet low food prices and hence more calorie intake have clearly added to the problem. By 2003 the average American was consuming 2,757 calories daily, roughly 23 percent more than in 1970. Compared to 1900, American men today are nearly 50 pounds heavier (but also nearly 3 inches taller, and far more likely to live to age 85).

Some popular writers such as the journalist Michael Pollan have tried to link the sometimes damaging cheapness of food in America to the artificial production incentives found in federal farm subsidy programs (Pollan 2006). Federal commodity programs do make some products cheaper for consumers, especially grain products, but the overall effect is a mixed one, since commodity programs at the same time have made sugar and dairy products more expensive. Economic analysis shows science-based productivity is far more important than commodity programs as an explanation for the cheapness of food in America today (Alston, Sumner, and Vosti 2006).

In Europe, historical concerns about food shortage have also been replaced now by anxieties about food excess. In 2006 the head of Europe's Directorate for Health and Consumer Affairs announced that obesity had actually become a bigger killer in Europe than tobacco (EU Food Law 2006). In this altered cost and health environment, amid such food abundance, it is understandable that the importance of pursuing still more agricultural science would decline. All the more so because there are several other reasons, in the minds of nonfarmers, to fall out of love with agricultural science. Many now oppose agricultural science on environmental grounds; others dislike farm science because they associate it with an industrial farming model dominated by corporate agribusiness; and still others dislike farm science because it gives them foods they would rather not eat.

Environmental Objections

The environmental case against science-based farming was made to greatest effect by Rachel Carson in her 1962 classic, *Silent Spring*, a book that documented the damage done to nature and people by the use in farming of synthetic insecticides such as DDT (Carson 1962). Carson, an aquatic biologist, was a literary celebrity who had already won the National Book Award ten years earlier for *The Sea Around Us*, which—like *Silent Spring*—reached a wide and influential audience through serialization in *The New Yorker*. In *Silent Spring* Carson focused on the damage to wildlife done by agricultural chemicals and the growing resistance of pest populations to those chemicals, yet her larger aim was to point to the human arrogance of science itself. In an opening epigraph she quoted E. B. White: "I am pessimistic about the human race because it is too ingenious for its own good." Carson's book had a large and lasting impact; it led not only to a specific ban on the production of DDT in America, but also to the formation of a broad and powerful national environmental movement and eventually passage in 1970 of a National Environmental Policy Act that created the Environmental Protection Agency (Lear 2002).

Carson's charge against excessive chemical use in agriculture was powerful and irrefutable, and eventually persuasive even to most commercial farmers, who had an interest in reducing toxic chemical use for pocketbook reasons (the chemicals are expensive), for occupational safety reasons (farmers are those most exposed to risk), and also for compelling agronomic reasons (insect populations can become resistant to the chemicals). Yet the technical strategy farmers preferred wasn't to abandon science in crop protection, but instead to use still more science. Industry developed newer chemicals with fewer harmful side effects and high-precision application equipment that reduced excessive or unwanted exposure. Farmers also cut chemical use by purchasing crop varieties bred scientifically for increased resistance to disease or insects. Among nonfarmers, however, Carson's skepticism about science led to a paradigm shift in public thinking about agriculture. The search began for an alternative to the highly specialized and chemically dependent style of farming that modern science had fostered. Social advocates for "alternative" farming systems that were less dependent on chemical or biological science began to proliferate.

One early leader in this effort was Wes Jackson, originally a nonfarmer, a Ph.D. geneticist, and creator of the nation's first environmental studies program at California State University. Jackson returned to his native Kansas in 1976 to found the Land Institute, where he promoted using polycultures of perennial crops rather than monocultures of annual crops. The philosophy was pure Rachel Carson: do not try to engineer or dominate nature; instead use nature as a model. In 1980 Jackson began employing the term "sustainable agriculture" to describe this alternative approach, and in 1985, in response to lobbying efforts in Congress from a growing list of like-minded nonprofit organizations, the Department of Agriculture finally initiated a program to promote Low-Impact Sustainable Agriculture, or LISA. Conventional commercial farmers in America wanted nothing to do with this LISA approach, ridiculing it as low *income* sustainable agriculture, but the idea gained traction, supported both by nonfarmers and by a small but culturally influential cohort of newer small farmers from the "back to the land" movement that had emerged during the countercultural rebellion of the 1960s and 1970s.

Not all modern environmentalists disapprove of highly specialized, technologically modern farming systems. James Lovelock, an independent scientist and the originator of Gaia theory in 1972, argues that an adequate protection of nature in some places may require a more complete domination of nature somewhere else. Lovelock would prefer to divide the landscape into three parts: one-third would be given over to industry, then one-third to highly *intensive* farming, allowing the final third to be left alone to evolve entirely free of human interference (Lovelock 2006). Lovelock's vision is fanciful, but it makes an important point. A low-input model of agriculture designed to imitate nature would—exactly like nature itself—produce less human food per hectare, thus requiring the use of much more land in farming. Thanks to science-intensive farming in America since 1950, the dollar value of total output has increased more than 100 percent even while the total land farmed declined by 25 percent. The total land area devoted to farming in America has actually declined on a per capita basis by more than 50 percent since 1920, thanks to high-input, science-intensive farming (Akst 2003). Carsonian environmentalists cannot refute this logic, but they resist accepting it because it requires them to endorse a larger rather than a smaller role for modern science.

Anti-Industrial Objections

A second source of resistance to science-based farming in rich countries, again coming mostly from nonfarmers, is social and cultural. It is born of a fear that more agricultural science will only lead to a more complete industrialization of farming, concentrated in the hands of corporate agribusiness. The cultural loss associated with applications of science to farming has been real; close agrarian communities built around small family farms have given way to a rootless culture of corporate agribusiness, box-store shopping outlets, and sprawling condominium construction. In Europe no less than in America, this loss of agrarian culture in the twentieth century remains a source of both personal pain and social regret.

In Great Britain between 1945 and 2005 the size of the agricultural workforce decreased by three-quarters and the total number of farms fell by nearly two-thirds. The one million or so families that left the land during this period typically found more profitable work options in the urban economy, but they and their children retained strong emotional attachments to farmstead and village and many have never felt fully whole in their new lives. In an emotionally candid memoir, Richard Benson describes the heartbreak he felt when his family's proud 200-year history of farming in Yorkshire came to an end in the 1990s as small, diversified pig farms lost out to larger and more specialized operations. Benson had made the sensible decision to leave farming for an attractive job in London as a magazine editor, but this did not lessen the pain of watching his parents' farm auctioned to developers, or the dismay at then seeing their traditional white chalk barn turned into a cute residential cottage with pastel flowers painted on every available interior surface (Benson 2006).

In America even those with no recent family connections to farming can legitimately regret how science-intensive agriculture has changed or eliminated small family farms, an evocative symbol of the spirit of free men and women since Thomas Jefferson's time. On small family farms, individuals can work for themselves rather than for a boss; they can take pride in providing for their families through the energy, skill, and mindfulness they bring to the handling of multiple seasonal tasks. The science-driven modernization of agriculture has rendered useless the skills of these traditional workers and taken away much of their precious independence. Most commercial farmers in America today still own their land, buildings, and equipment, but increasingly they now rely on large food-processing companies to tell them what to produce, and how. More than 80 percent of poultry and egg production and more than half of all hog and fruit production in the United States now take place under contracts in which the processors downstream dictate which feed mixes or genetic strains farmers must use upstream.

Landscape changes associated with industrialized farming are also culturally jarring and unattractive to nonfarmers who enjoy the countryside as weekend tourists or owners of second homes. Affluent urbanites in the United States and Europe enjoy visits to the country so much that in recent decades a lucrative industry called "agritourism" has sprung up. City dwellers now pay as much as $240 a night for the privilege of spending time on a traditional Vermont dairy farm, exposing their children to such bucolic pleasures as feeding cows and gathering eggs. What the agritourists and country homeowners expect to encounter when they exit the city is a traditional landscape featuring stone buildings, bales of hay, flocks of sheep, and grazing dairy cows. They do not want to see the landscape of modern agribusiness: mile after mile of monocropped fields with little evidence of human settlement, except for a few large and suspiciously windowless aluminum buildings surrounded by foul-smelling lagoons of animal waste.

A significant cultural backlash against industrialized agriculture began in America in the 1960s, in tandem with the Carsonian environmental backlash. The poetic voice of this movement was Wendell Berry, a nonfarmer who had studied creative writing at Stanford and worked for a time as a teacher of English at New York University. Berry returned to his home state of Kentucky in 1965 to write, operating a small farm in his spare time. In his writing Berry stressed the importance of using technologies that were locally producible, easily repaired on the farm, not dependent on fossil fuels, and not likely to change the nature of the community; he admired the nineteenth-century farming techniques of the Amish and made a point of using horses when he plowed his own fields (Zencey 2002). Much in the manner of the original Romantic poets, who lamented industrial growth in their time, Berry lamented the twentieth-century industrialization of agriculture.

A more angry, populist rejection of science-based farming in the United States followed in the 1970s from Jim Hightower, a talented, working-class Texan who eventually became nationally known as an anticorporate broadcaster and humorist. In a 1973 book he titled *Hard Tomatoes, Hard Times,* Hightower exhaustively documented the links he saw between publicly funded agricultural science at the nation's land-grant colleges and the demise of small farms, the displacement of farm workers, and a consolidation of the corporate power of agribusiness. In his attack on America's land-grant research system, Hightower had found a ripe target; the university scientists in this system were so pleased with the productivity enhancements they were spawning for their commercial and corporate sponsors that they never saw the populist attack coming. As one land-grant administrator said at the time, "We were crucified with our own data" (McCalla 1974). Following Hightower's effective attack, it became far more difficult for agricultural scientists to claim an entitlement to taxpayer money for a next generation of farm productivity enhancements.

Consumer Objections

A third strain of social resistance to agricultural science comes from quality-conscious food consumers. The science-intensive farming systems that rose to prominence in America and Europe during the middle years of the twentieth century provided low cost, variety, and great convenience, but very little of what many consumers would consider quality. When affluent and well-educated consumers in today's rich countries confront the foods they see emerging from their modern, industrialized farming systems, the response of many is that they deserve better.

To anyone who respects traditional cuisine, the food products grown on factory farms today and then processed and marketed through modern food industries are indeed a rude affront. They are conceived and engineered by corporate "food scientists" pursuing dubious objectives, such as glossy color or extended shelf-life. It is modern food science that gives us enzymes to enhance crispness or spreadability, innovative flavor and aroma technologies, improved canning systems, and a constantly expanding repertoire of new processing technologies such as electric fields, high hydrostatic pressure, light pulses, and irradiation. The crops and animals raised on commercial farms today are designed as much to withstand these abusive downstream processing and packaging treatments as to ensure traditional consumer benefits such as freshness or nourishment. Quality-conscious citizens have understandably been repelled, and since the 1960s many have begun to demand foods with fewer additives, foods grown with fewer chemicals, and locally grown foods that are more likely to be fresh.

Consumers in rich countries today also want foods grown in what they consider an ethical manner. In the United Kingdom, ethical labeling of foods now allows consumers to "vote with their shopping trolley" every time they move down the supermarket aisle. They will find foods labeled organic (grown without synthetic chemical fertilizers or pesticides), or Soil Association certified (subcategory of organic), or Fairtrade (produced by farmers accredited by the Fairtrade Foundation), or free range (from animals that lived outdoors), or Linking Environment and Farming (LEAF) (produced by farmers committed to improving the environment), or according to Carbon Footprint (shows weight of carbon produced in manufacture of the product), or marked with an airplane (shows products flown into the United Kingdom), or Rainforest Alliance (produced in compliance with Rainforest Alliance guidelines), or Utz certified (coffee products grown in a socially and environmentally friendly fashion, according to Utz Kapeh certifiers), or Vegetarian Society approved (free of animal products).

In the United States the consumer rebellion against industrialized and "unethical" food first gained a broad base during the countercultural movement of the late 1960s when it became fashionable to spurn corporate food products and replace the consumption of processed foods with more natural alternatives such as brown rice, buckwheat, millet, couscous, bulgur, unpearled barley, and rye berries. Warren Belasco recalls that the first rule of the day was, "Don't eat anything you can't pronounce," which meant no propylene glycolalginate in the bottled salad dressing (Belasco 1989, p. 40). This reification of what is "natural" is in part a cultural reaction to the hegemonic expansion of modern science. Advances in modern science tend to diminish both unspoiled nature and unquestioned faith, prompting those with a strong romantic or spiritual side to register their objections by seeking foods that incorporate less modern science.

This has in fact been happening in rich countries for a long time. In America, efforts to promote the consumption of nonindustrial foods began in 1829 when a Connecticut clergyman, Sylvester Graham, invented "Graham bread" (ancestor of the modern graham cracker), which contained none of the chemical additives that bakeries had begun using to increase whiteness. Graham was also a patriarch and a prude: he thought women should return to milling their own flour and believed in vegetarianism as a means to help control sexual passions (Fromartz 2006). The modern organic food movement also has early anti-industrial roots. The first strong proponent of farming without synthetic chemicals was an Austrian philosopher named Rudolf Steiner (1861–1925), who founded a vitalist school of spiritualism called anthroposophy. Vitalism was the once-dominant view that living things had a chemical composition different from nonliving things. This belief was exposed as erroneous by modern chemical science after 1780, yet Steiner and his followers persisted well into the twentieth century making assertions that plant life could thrive only on the recycled manure of living animals. Steiner rejected the applications of modern chemical science then sweeping through European agriculture, arguing that traditional peasant practices were "far wiser" than the innovations introduced by scientists (Silver 2006). A thorough mystic, Steiner also believed in cosmic rhythms, human reincarnation, and the lost city of Atlantis. Near the end of his life in 1924 he began advocating what he called "biodynamic" (life-force) agriculture, which—among other things—advocated compost preparations that included chamomile blossoms and oak bark, and forbade the use of synthetic nitrogen fertilizers.

As the twentieth century moved ahead, this urge to resist science-based farming next spread in England following the publication in 1940 of a book by Sir Albert Howard titled *An Agricultural Testament.* Howard, who had worked in India and was influenced by Eastern spiritual concepts, asserted that food produced with synthetic nitrogen fertilizers would leave people physically stunted and prone to disease: "Artificial manures lead inevitably to artificial nutrition, artificial food, artificial animals, and finally to artificial men and women" (Shapin 2006). Howard's ideas were picked up in Lady Eve Balfour's 1943 book, *The Living Soil,* which inspired the creation of the Soil Association in 1946, which is still the institutional guardian of organic farming traditions in Great Britain. His Royal Highness the Prince of Wales is today an avid organic farmer, a leading patron of the Soil Association, and the most prominent current exemplar of this blue-blood attachment in England to preindustrial, chemical-free agriculture.

Howard's influential thinking also traveled to the United States, where it had an impact on Jerome Irving (J. I.) Rodale (1898–1971), an accountant who grew up on the Lower East Side of Manhattan. Rodale read Howard's work and concluded the idea made sense: "Surely the way food is grown has something to do with nutritional quality"

(Fromartz 2006). Rodale then coined the term "organic" farming, and in 1942 began a new career as publisher of his own magazine, *Organic Gardening and Farming.* Rodale also became committed to alternative methods of health care, promoted through his founding of *Prevention* magazine in the 1950s. In his personal life Rodale rejected white flour (like Graham), did not use plastic or aluminum utensils, and took more than 100 nutritional supplements daily. In 1945 Rodale even beat Rachel Carson to the punch by warning about the dangers of DDT.

For several decades Rodale's organic-farming magazine had only a small readership, mostly among suburban backyard gardeners. The popularity of organic farming in America only began to spread significantly following the rise in consumer affluence in the 1960s, the popularity of environmental and countercultural concerns in the 1970s, and then increased media attention given to farm insecticide scares in the 1980s and 1990s. Organic products were at first not widely available beyond specialty markets or health stores, which did almost as much to hold down consumer interest as the fact that they cost 10–40 percent more than conventionally grown products. A major boost was provided in 1990 when Congress mandated the creation of a single national standard for the marketing of organically grown products. Once this uniform standard was established, consumer confidence increased, and commercial sales of certified organic products in the United States began growing at annual rates of roughly 20 percent, to reach $13.8 billion by 2005.

Despite such impressive growth, the organic sector remains small in the United States, making up only 2 percent of total U.S. food purchases and using only four-tenths of one percent of U.S. cropland (Greene 2006). Dairy has been one of the fastest-growing segments of the organic food industry in the United States, yet as of 2003 only 1–2 percent of all dairy cattle in California and Wisconsin, the two top states for organic milk production, were being sustained on organic feed. Organic foods and organic farming are more popular in Europe, in part because governmental programs explicitly promote them. The share of agricultural land under organic production in Europe has now reached 4 percent, double the level of 1998, and many governments in the European Union have established even higher targets for organic production. In 2001 the EU-15 collectively spent $559 million in cash payments to reward farmers who converted to organic methods (Dimitri and Oberholtzer 2006).

When consumers in rich countries pay more for organically grown foods, what do they get in return? Retailing surveys show 70 percent of U.S. consumers today buy organic foods to avoid pesticide residues, and in Europe as well, personal health is the key consumer motivation. Yet regulatory systems in both the United States and the European Union have now placed such strict limits on the pesticide residues permitted on conventional foods that the added advantage gained from purchasing organic is surprisingly small. In 2003 the Food and Drug Administration analyzed several thousand samples of domestic and imported foods in the U.S. marketplace and found that only four-tenths of one percent of the domestic samples and only one-half of one percent of the imported samples had detectable chemical residues that exceeded tolerance levels.

Exposures to pesticide residues on conventionally grown foods in the United States appear to be safe, and by a wide margin according to prevailing international standards. Studies of total diets by the FDA have measured the *highest* average daily intake of thirty-eight different pesticides for different population subgroups, and then compared these measures with the United Nations FAO/WHO "acceptable daily intake" (ADI), a level that in turn is quite conservative (a typical human exposure at one percent of the ADI level constitutes an exposure *ten thousand times lower* than a level that still does not cause toxicity in animals). The FDA found in its comparison that for all but four of the thirty-eight pesticides it studied, estimated exposures in the United States are now at less than one percent of the ADI value, and for the other four, all are below 5 percent of the ADI value. Carl K. Winter and Sarah F. Davis, food scientists with the University of California–Davis and the Institute of Food Technologists, conclude from these data, "[T]he marginal benefits of reducing human exposure to pesticides in the diet through increased consumption of organic produce appear to be insignificant" (Winter and Davis 2006).

In Europe regulatory limits on pesticide residues for conventional foods also seem to be working well enough. Government monitoring programs have indicated that residue levels in foods are typically only a small fraction of the established safety level, and the EU Commission in 2000 found that only 4 percent of foods sampled contained residues above this safe legal limit. As for other possible concerns about conventional foods versus organically grown foods, including nutrient qualities, nitrate levels, or naturally occurring toxins, organic advocates frequently make claims of benefit, but the science community is skeptical. Winter and Davis find that the available data are inconclusive: "[I]t is premature to conclude that either food system is superior to the other with respect to safety or nutritional composition" (Winter and Davis 2006). Claire Williamson from the British Nutrition Foundation reaches the same conclusion: "From a nutritional perspective, there is currently not enough evidence to recommend organic foods over conventionally produced foods" (Williamson 2007).

The environmental benefits claimed by organic farming are also challenged by the scientific community. Because organic farmers are not permitted to use synthetic chemical fertilizers, they must replace soil nutrients by using more land for animal pasture, and for rotations of cover crops that they plow under as "green manure." Compared to conventional farming, the result is a larger spatial footprint on the land per unit of production, a factor offsetting some of the environmental gains that might come from restrictions on synthetic pesticide use. Because of restrictions on the use of chemical fertilizers, any serious effort to scale up organic farming would necessitate expanding the area of land devoted to farming, meaning more rather than less environmental disruption in the countryside (Smil 1997). Prohibitions in organic farming against the use of all synthetic herbicides can also have adverse environmental impacts, because they block the use of modern no-till farming practices rated as superior to organic farming along all environmental criteria (Trewavas 2004).

Scaling-up difficulties linked to land requirements plus the persistently higher production costs of organic methods imply that organic farming is never likely to constitute more than a small part of commercial agriculture in rich countries. In the United States the state of Maine is home to the largest state-level organic association in the Untied States, with more than 5,200 members, yet certified organic farming still makes up less than 5 percent of all farms in the state and only 4 percent of farmland (MOFGA 2006). Moreover, where organic farming has been able to scale up, as in the dairy and vegetable sectors, it has usually required the adoption of industrial-size production systems similar to those now found in conventional commercial farming, undercutting consumer support. Most organic milk now comes from specialized dairy operations handling hundreds of animals within confined feed and milking facilities, and most organic vegetables today come from giant corporate farms that ship over long distances and sell through supermarkets. In 2002 in the United States only 13 percent of organic vegetable sales were still being made by small farmers through local farmers' markets (Fromartz 2006). Organic purists have naturally been offended by this trend. The last straw for some came in 2006 when Wal-Mart, the mass-marketing chain, announced it would start offering more organic foods, a move that then helped pull a number of commercial brands such as Pepsi, Rice Krispies, and even Kraft Macaroni & Cheese into the organic game. Leading guardians of the organic movement, such as the Cornucopia Institute, complained that Wal-Mart was cheapening the value of the organic label, in part because it sourced some of its organic products from factory farms in China (FTR 2006).

With the organic standard no longer pure enough, what are quality-conscious food consumers to do? In 2006 a best-selling book by Michael Pollan titled *The Omnivore's Dilemma* skewered the organic movement's new drift toward industrial, corporate-sized operations and fantasized about disconnecting from industrialized food entirely by returning to localized food economies, home gardening, and even hunting and gathering (Pollan 2006). A growing international Slow Food movement, founded in Italy in 1986, has emerged as another way to fight back against food industrialization. Bolstered by concerns that shipping food from a distance will mean needless fossil-fuel consumption and hence increased global warming, a movement promoting locally grown and purchased food has more recently taken off. To accommodate enthusiasm for local food, residential real-estate developers in the United States who once built their condo complexes around golf courses are now building them around working farms. In suburban South Burlington, Vermont, 334 homes are being built to surround a forty-acre farm that will grow corn and organic produce. In Albermarle County near Charlottesville, Virginia, a new 2,300-acre local food community called Bundoran Farm will feature a cattle ranch with riding trails, two ponds, an apple orchard, and home sites for $400,000 and up (Munoz 2007).

Downgrading Agricultural Science

All of these recent cultural shifts have tended to weaken the reputation of agricultural science. Food purists, environmentalists, populists, and agrarian romantics all agree that agricultural science is a large part of the problem and not likely to be part of the solution. They have no serious plans to return to hunting and gathering any time soon, but neither do they want more of their tax money invested in still more productivity-boosting agricultural science, since they suspect this will take them even further down a road they would rather not travel. Governments in rich countries have become aware of this new, popular turn against agricultural science and have responded accordingly. Since the 1980s public investments in farm science have stalled in rich countries, and in some cases even declined. In one sample of twenty-two rich countries, real public spending on agricultural research and development (R&D) reversed its traditional upward path after the 1980s and began declining at an annual rate of 0.6 percent in the 1990s (Pardey et al. 2006).

The United States has not reduced public spending on agricultural research overall, but it has cut the share of that research that goes for further on-farm productivity growth. Back in 1940 a full 40 percent of all federal R&D spending in the United States ($29 million out of $74 million) went to agricultural research, either at the USDA or at state agricultural experiment stations (Fuglie et al. 1996). Following World War II, the federal government took on a new set of demanding research tasks (atomic energy, cold war military security, and the space race), yet investments in agricultural R&D continued to expand in real terms at a strong 2.8 percent a year between 1950 and 1970. Following Jim Hightower's 1973 critique, however, federal funding for agricultural research stopped growing and the research focus moved conspicuously away from providing new productivity gains. Agribusiness interests continued to drive priorities, but the focus turned to downstream food processing, food safety, and export promotion. The environmental community also got more research money for LISA, and the agrarian populists got more for quality-of-life enhancement in rural communities. By 1992 nearly one-third of all public agricultural research spending in the United States was earmarked for purposes other than augmenting productivity (Alston, Pardey, and Smith 1998; Fuglie et al. 1996).

Traditional agricultural scientists in the United States complained, but they were told by the nation's political leaders to take their medicine. A 1992 Office of Technology Assessment report from the U.S. Congress titled *A New Technological Era for American Agriculture* echoed Rachel Carson and Jim Hightower in concluding, "The ability of the land-grant system to carry out its historic missions is becoming increasingly suspect . . . Public interest groups have become increasingly critical of the emphasis on productivity in agricultural research" (OTA 1992, p. 411). As a consequence, between 1988 and 2000 overall, federal funding through the land-grant system for academic agricultural research, teaching, and extension declined by 8 percent in constant dollars.

Frustrated agriculturalists accused political leaders of losing faith in the social benefits of science, but public science investments in fields other than agriculture remained on a strong upward trajectory. It was only agricultural science that was out of favor. During the first five years of the twenty-first century, while federal agricultural research spending remained flat at approximately $2.7 billion per year, congressional appropriations for medical research spending through the National Institutes of Health (NIH) increased from $18 billion to $28 billion. By 2007 NIH was outspending USDA on research by fifteen to one.

In Europe, public spending for agricultural science met a similar fate, worsened by a growing frustration with the high public cost of storing and disposing of surplus agricultural production, a political rise of green parties opposed to science-intensive farming, and (especially in the post-Thatcher United Kingdom) a shift away from relying on public-sector funding for anything. Public spending for farm research in the United

Kingdom began falling at an annual rate of negative 0.2 percent in the 1980s (Alston, Pardey, and Smith 1998). In 1987 Prime Minister Margaret Thatcher even sold off to Unilever the crop breeding programs and the farm site of the U.K. Plant Breeding Institute at Cambridge. In both Europe and the United States from the 1980s onward, market fundamentalists embraced a hope that any needed farm research would be provided by the private sector.

This hope made considerable sense in the United States, where strong intellectual property protections for biological innovations provided adequate R&D investment incentives to corporate labs. In the 1980s the U.S. Supreme Court authorized standard utility patents first for microorganisms and then for plants and animals, and as a result, private-sector R&D investments in plant breeding quickly rose to more than twice the investment level of the faltering public sector (Frey 1996), and total private spending on all agricultural R&D soon increased to exceed total public-sector spending (Klotz-Ingram and Day-Rubenstein 1999). By an accident of timing this shift took place just as genetic engineering techniques were becoming available to crop science, explaining why this new technology emerged in the end mostly as a private- rather than a public-sector project.

This also helps to explain why agricultural GMOs would become such an easy target for the opponents of agricultural science. The first generation of GM crops that emerged from the corporate labs—crops engineered to reduce the cost to farmers of insect and weed control—were distinctly unappealing to groups already mobilized against farm science. Agrarian romantics and populists were offended because the technology was being developed by profit-making multinational corporations primarily for the benefit of seed companies and large commercial farmers, and also because in rich countries it would be sold with use restrictions that obliged farmers to purchase seeds anew every year, further compromising their traditional independence. Carsonian environmentalists were offended because gene transfer was so clearly an attempt to engineer and dominate nature rather than live within nature's normal reproductive constraints, and because some of the new GM seeds produced their own pesticides. Quality-conscious food consumers were offended most of all. They were alarmed to see food companies introducing onto supermarket shelves without a warning label something consumers had never asked for, foods that contained "foreign genes."

Withdrawing Support for Agricultural Science at Home—and Abroad?

We have shown in this chapter how successive applications of modern science to farming in the twentieth century eventually led, in rich countries, to a lower need and a reduced desire for still more agricultural science investments, causing the governments of rich countries in the end to cut back on such investments. When governments in rich countries decided to downgrade agricultural science in this fashion, little was lost. Their farmers were already highly productive and food was abundant and cheap, especially relative to income. But these rich countries not only cut back on public support for agricultural science at home; in the chapter that follows it will be seen that they also cut back on their support for agricultural development and agricultural science abroad. In the developing countries of Asia and Latin America, this had little impact, because these countries had long since built up an independent capacity to finance their own investments in agricultural science. For the aid-dependent countries of Africa, however, the turn in rich countries away from science-based agricultural development brought serious local consequences. Turning away from scientific innovation was an affluent taste that Africa's poor farmers could not yet afford.

3

Withdrawing Support for Agricultural Science in Africa

Farming in Africa is a world apart from farming in Europe or North America. Agriculture is the main source of income for roughly two-thirds of all Africans, compared to fewer than 5 percent of those in Europe or the United States. In Africa it is small farms that dominate, not large farms; 80 percent of farms are smaller than two hectares in size. Also, 90 percent of farmers practice diversified crop production rather than specialized monocropping. In Africa crop yields remain extremely low. In the United States in 2005, farmers planting maize harvested an average of 9.3 tons per hectare of land; in Kenya, maize farmers harvested only 1.6 tons per hectare and in Malawi only .8 tons (FAOSTAT 2007). Chemical use is extremely low. In the industrial world as a whole, fertilizer use now averages 117 kg per hectare of arable land; in sub-Saharan Africa average fertilizer use is only 9 kg per hectare, and for 23 of the poorest countries in the region less than 5 kg per hectare (Camara and Heinemann 2006). Mechanization is nearly nonexistent. In the United Kingdom there are 883 tractors per 1,000 agricultural workers, whereas in sub-Saharan Africa there are now two per 1,000, which is actually a drop of 50 percent from the 1980 level of three.

Income is the biggest difference between science-poor farming in Africa and science-rich farming in Europe and North America. In France in 1998, annual value added per agricultural worker was $37,000 and rising; in sub-Saharan Africa that year it was just one one-hundredth as much on average ($379 per year) and actually falling (World Bank 2001). In the United States the average yearly income of farm households now exceeds that of nonfarmers, and the average net worth of American farm households in 2003 was $664,000 (Offutt and Gundersen 2005). In Kenya in 2000 the average yearly income of a farming household was just $553. In Zambia, $122. In Mozambique, just $59 (Jayne, Mather, and Mghenyi 2005). Nearly 80 percent of all those in Africa officially classified as poor are farmers.

The typical African farmer today is a woman (60–80 percent of farm labor in Africa is provided by women) who works most of the daylight hours preparing meals and raising children in addition to tending crops. She cannot read or write and lives in a dwelling made of sticks, mud, and thatch, with no plumbing or electricity. Water and wood for cooking must be carried in. She keeps some goats, grows some vegetables in a garden area close to her home, and tends small plots of land a short walk away, where she plants food crops such as yams, maize, beans, sorghum, millet, or cassava, and perhaps also some cotton for cash income. In the years when the rains are good and these crops do well, she will be able to meet her family's immediate needs and have a small surplus to be sold or bartered locally. Her fields have no irrigation, so if the seasonal rains fail, the crops will also fail. She uses little or no chemical fertilizer because it is expensive, because she lacks access to credit for purchases prior to the harvest, and because using fertilizer will be a waste of money if the rains fail. Her children often join her in the fields, either strapped to her back as infants or, once they can walk, kept busy tending goats, shooing birds, and pulling weeds. African women farmers are hardworking, skillful, and highly resourceful. Precisely because they are poor they cannot afford to waste any time, labor, or materials. Yet because their minimal tools, seeds, and input supplies are so limiting, even the most persistent efforts bring little reward. In the memorable phrase of T. W. Schultz, they are "efficient but poor" (Schultz 1964).

In Africa, in other words, farmers today are not engaged in specialized factory farming. They are planting heirloom varieties in polycultures rather than scientifically improved varieties in monoculture. They have a food system that is traditional, local, nonindustrial, and very slow. Using few purchased inputs, they are *de facto* organic. And as a consequence they remain poor and poorly fed.

While applications of modern science were increasing the productivity of farmers in Europe, North America, and Asia during the nineteenth and twentieth centuries, farmers in Africa were mostly left behind. Colonizing Europeans invested in research to improve some of the specialty crops Africans were obliged to grow for export—such as cocoa, coffee, tea, cotton, peanuts, and sugar—but traditional African food crops such as millet, cassava, yams, and sorghum remained mostly unimproved by scientific plant breeding. The tools African farmers used continued to be simple hand hoes, planting sticks, machetes, and wooden plows. The only power they had came from their own physical labor or that of some animals. Even today, nearly all farm household travel is by foot (Calvo 1998).

Even compared to other developing regions, African farming stands out as dramatically nonproductive. The accompanying table, assembled by Jeffrey Sachs at Columbia University and the United Nations (UN) Millennium Project, provides comparative measures of farm technology uptake and productivity in sub-Saharan Africa, Asia, Latin America, and North Africa and the Middle East.

The table indicates how far Africa has fallen behind other developing regions in the uptake of modern agricultural science. The "modern va-

e 3.1. Agricultural Technology and Productivity by Developing Region

eloping on	Share of area planted to modern varieties (percent)		Cereal yield (metric tons per hectare) 2000	Average annual growth in cereal yield, 1980–2000 (percent)	Average annual growth in food production per capita, 1980–2000 (percent)
	1970	1998			
-Saharan ca	1	27	1.1	0.7	−0.01
	13	82	3.7	2.3	2.3
n America	8	52	2.8	1.9	0.9
dle East North ca	4	58	2.7	1.2	1.0

rce: Adapted from Sachs et al. (2004), based on Evenson and Gollin (2003); World Bank (2003); FAOSTAT abase.

rieties" of crops mentioned in this table are not GMOs; they are just hybrid varieties or the conventionally developed, high-yielding varieties of the Green Revolution of the 1960s and 1970s. It is in part because so little area in Africa is planted to these improved crops that yields remain so low. Africa's growth in yields from this low starting point has been so slow that the absolute yield gap with the rest of the developing world has continued to widen. The ominous final result, as seen in the last column of the table, has been a *negative* per capita growth rate of food production in Africa.

These science-starved farming conditions in Africa should be motivating outsiders to support much larger local investments in agricultural research. Local research investments are critical in Africa because this region—due to its distinct crops and ecological conditions—is less likely than any other to benefit from "spillovers" of farming technologies developed elsewhere. Philip Pardey, director of the University of Minnesota's International Science and Technology Practice and Policy Center, has estimated that Africa is only about 20 percent as likely to benefit from international farm research spillovers as the rest of the developing world (World Bank 2008). Yet in recent years local investments in agricultural science in Africa have faltered, partly because international assistance to agriculture has declined.

In this chapter I contrast Africa's need for a science-based productivity upgrade in farming to a curious failure on the part of African governments, in recent decades, to invest in this essential project. I trace a significant part of this recent policy failure in Africa to international sources, including a sudden withdrawal of international assistance to agriculture in Africa beginning in the 1980s, which was worsened by a surprising hostility to scientific advances in farming among some newly influential members of the international NGO community working in Africa. These significant moves against the promotion of science-based agricultural development in Africa all predated the controversy over GMOs. In fact, they help explain that later controversy. Well before GM crops came along in the 1990s, strong international winds were already blowing Africa away from the science-intensive path to farm productivity that other developing nations had followed.

The Pro-Poor Potential of Agricultural Science in Africa

Many who accept modern science as good for agricultural productivity overall will nonetheless question its potential to benefit the poor. In Latin America, they point out, modern agricultural science has had a history of worsening the circumstances of the rural poor. When the development of chemical insecticides suddenly made cotton production profitable in Central America after the Second World War, one result was an increase in the commercial value of land. The poor were then pushed off the land by the landlord to make way for cotton production (Williams 1986). Where the rural poor lack secure access to land, as across so much of Latin America, it does become hard to find a new farm technology capable of being both pro-growth and pro-poor.

Fortunately in Asia and Africa, the poor generally have more secure access to agricultural land. Whereas in Latin America there are eighty-two landless people in the countryside for every one hundred smallholder farmers, in Asia there are only fifty-three, and in Africa only fifteen (Hazell and Haddad 2001). This greater prevalence of land-secure smallholder farmers among the poor in rural Africa increases the chance that they will benefit from a farm-technology upgrade. Yet not just any upgrade will do. A new farming technology will be pro-poor as well as pro-growth only if it raises the total factor productivity of small as well as large farms. If a new technology gives a productivity boost only to the largest farms, the result can be a fall in crop prices that hurts those who produce food on smaller farms. New farming technologies will also be more likely to deliver productivity gains to the poor on small farms if they reduce rather than increase weather risks, if they target disadvantaged areas where the poorest small farms operate, and if they target the staple food crops currently grown by such farmers (Meinzen-Dick et al. 2004).

Michael Lipton has added some important refinements to these generalizations. He shows that when agricultural land is becoming increasingly scarce relative to population, which is now the case in some densely settled parts of Africa, technologies will be pro-poor only if they raise the productivity of land and water *faster* than they raise the productivity of farm labor, otherwise demand for hired farm labor may lag, slowing income growth among the landless rural poor. A technology like a new chemical herbicide might fail this test, as it does more to increase the productivity of labor (less labor needed to pull weeds) than it does to increase yield per hectare. As a second test, the rise in total factor productivity in farming must be great enough to protect farmers from the ensuing reduction in food prices as production increases, otherwise poor farmers will lose in lower prices for their commercial crops more than they are gaining in lower costs (Lipton 2005). Africa's weak rural infrastructure and marketing institutions can get in the way of a favorable outcome here, since sudden seasonal surges in on-farm productivity can cause local food gluts that collapse market prices.

To date little productivity growth of any kind—pro-poor or otherwise—has been brought to farms in Africa. Measures are hard to agree on because of the sketchy quality of most official production data, exchange rate and rainfall fluctuations, and estimation problems associated with fluctuating household labor inputs. In addition, statisticians in Africa must account for a multiplicity of unfamiliar crops and animal products, many of which are never marketed. Some years ago Steven Block overcame some of these limitations by looking at the 1963–88 period as five separate five-year intervals and by translating all farm production into "wheat unit" equivalents. When he did this he found that productivity in African farming has been uneven but low overall, with growth rates as high as 1.4 percent and 1.6 percent for the first and last of these five-year intervals, but growth rates in two of the other three intervals that were negative (Block 1994). The approach used by the World Bank has been to measure changes in value added per farm worker over time, an approach that again reveals Africa's uneven and weak performance. While value added per farm worker in Asia was increasing rapidly under the influence of the Green Revolution between 1980 and 1997, in Africa annual value added per farm worker was actually declining, from $418 to $379 (World Bank 2001). It is because agricultural labor generates so little economic value in Africa—on average only about a dollar a day—that the personal income of Africans in the countryside remains so low.

Why so few productivity gains in Africa? Critics of science and critics of Africa say this is proof that agricultural research investments do not pay off there. The evidence suggests otherwise. Very few significant agricultural research investments have yet been made to benefit smallholder farmers in Africa, but when made they do tend to pay off. Colin Thirtle, Lin Lin, and Jenifer Piesse have calculated that the weighted average rate of return to agricultural R&D spending in Africa's farm sector has been a respectable 22 percent (Thirtle, Lin, and Piesse 2003). In its 2008 *World Development Report* the World Bank estimates, from a review of 188 different studies carried out in Africa between 1953 and 1997, that the average rate of return on agricultural research investments in Africa has been above 30 percent (World Bank 2008). Links between research investment and poverty reduction also can be strong in Africa, assuming the new technologies developed pass the tests mentioned above. Simulation studies in Ethiopia show that any productivity enhancement for staple-crop farmers secured through technology change will benefit farmers in even the poorest areas of the country (Hazell 2005). The problem with agricultural science in Africa has not been one of inadequate pay-off; the problem has been inadequate local investment, inadequate *pay-in*.

The agricultural research investments most needed in Africa will have to come from the public sector or from private foundations, because the staple food crops grown by Africa's poor remain of limited interest to private seed companies. The poor farmers who plant these crops are not good seed-buying customers, so corporate science pays little attention to their needs. In 2000 private agricultural R&D spending made up only 1.7 percent of total agricultural R&D spending in Africa, a minuscule share of a small total. A similar disinterest in the poor was shown by private companies at the outset of the Green Revolution in Asia, which is why governments and philanthropic foundations had to take the lead in funding and managing that technology upgrade. Governments in Africa have so far failed to play this essential role.

When the states of Africa gained formal independence in the 1960s, the expatriate European community departed and took much of Africa's modern farming knowledge with them. Local capacity was rebuilt somewhat in the 1970s through foreign aid transfers that did allow some leading states in Africa to gain a minimal local capacity in operating crop science labs, greenhouses, and field research stations. But then came the severe budget constraints of the 1980s and 1990s—brought on by unserviceable external debts and structural adjustment demands from international financial institutions—which forced cutbacks in public-sector spending across the board in Africa. New public spending for agricultural research stalled, and Africa's nascent national farm research systems never rose above a low capacity. The weakness of Africa's research investment performance is revealed in the accompanying table, which shows levels and trends in agricultural research spending between 1981 and 2000 for the developing countries as a whole, for sub-Saharan Africa in particular, and also for various other country groupings.

The table shows that from an already low starting point in 1981, Africa increased its public spending for agricultural research by only 14 percent during the 1980s, and then by only 7 percent in the 1990s. In contrast the developing countries of Asia—from a much higher starting point—increased their public spending on agricultural research by 59 percent in the 1980s and then by another 55 percent in the 1990s. In one sampling of twenty-seven countries in sub-Saharan Africa, public

Table 3.2. Public Agricultural R&D Spending (Millions of 2000 International Dollars)

	1981	1991	percent change 1981–1991	2000	percent change 1991–2000
All developing countries (117)	6,904	9,459	+37	12,819	+36
Sub-Saharan Africa (44)	1,196	1,365	+14	1,461	+7
Asia-Pacific developing countries (28)	3,047	4,847	+59	7,523	+55
China	1,049	1,733	+65	3,150	+82
India	533	1,004	+88	1,858	+85
All high-income countries (22)	8,293	10,534	+27	10,191	−3

Source: Adapted from Pardey et al. (2006), Table 2.

spending on agricultural R&D in half of these countries actually declined during the 1990s (World Bank 2008).

These discouraging figures should probably be adjusted further downward because they do not separate funds spent in Africa on actual scientific research from money spent for staff salaries. African governments have made the error of spending too much on salaries—hiring new and poorly trained scientists—and too little on actual research. In the sampling of twenty-seven African countries noted above, total numbers of agricultural research staff increased threefold between 1971 and 2000, but average spending *per scientist* fell by nearly half (Beintema and Stads 2004). When the recent growth of Africa's underfed population is taken into account, total agricultural R&D outlays on the continent are revealed even more clearly as inadequate. While the developing world as a whole was increasing its per-capita public spending on agricultural science by 30 percent between 1981 and 2000, per-capita spending in Africa actually fell by 27 percent (Pardey et al. 2006).

How much *should* the governments of Africa be investing in agricultural science? The International Food Policy Research Institute in Washington, D.C., has used a policy impact model to calculate what would be required for sub-Saharan Africa to replace its current "business as usual" policies in the agricultural sector with a revised set of "visionary" policies that might reduce numbers of malnourished children in the year 2025 from the 38 million projected under business as usual to only 9 million projected under the visionary approach. This study assumes that under business as usual all of sub-Saharan Africa will spend only $10.2 billion on national agricultural research between 1997 and 2025, whereas at least $15.0 billion would be required in the visionary scenario. The implication of this study is that sub-Saharan Africa is currently spending only two-thirds of what it minimally needs to spend on agricultural research to prevent a future of persistent poverty and hunger (Rosegrant et al. 2005). More than just bigger operating budgets will be needed, of course; investments will also have to be made in institutional capacity and human capital to revitalize science training within national university systems and to increase science literacy in government. Young African university students take an aggressive interest in science but lack access to both resources and institutional support.

One critical factor holding back agricultural science in Africa since the 1980s has been stagnant international donor support. The preceding table reveals that the only other countries to have faltered as badly as Africa in public funding for agricultural science in the 1980s and 1990s were high-income ones, where (as noted in the previous chapter) public spending for agricultural science slowed badly in the 1980s and then actually went into a decline in the 1990s. When the governments of these rich countries began cutting back on spending for agricultural science at home, they decided—with characteristic myopia—to de-emphasize agriculture and agricultural science abroad as well, which hit aid-dependent Africa hard. Governments of developing countries in Asia and Latin America were not harmed by the withdrawal of donor support for agricultural development that took place in the 1980s and 1990s because by then, thanks to the success of the Green Revolution, they were far less dependent on foreign aid. In Africa, unfortunately, when the donor community stopped supporting agricultural development, the governments of the region lacked the local resources to fill the gap.

Apart from donor influence, political leaders in all developing countries will face strong temptations to underinvest in scientific research, since their political time horizons will be short compared to the significant time lag (on average ten years in the case of agriculture) between making a research investment and getting the first tangible local payoff. R&D spending for agriculture is particularly easy for leaders in poor countries to neglect, since smallholder farmers in developing countries tend to be a weak political constituency. They may be numerous, yet they are physically distant from the capital city and less likely to pose a political threat compared to more vocal and better-organized city dwellers in the capital. Poor farmers in all developing countries are also more likely to be women (often politically disenfranchised), more likely to be illiterate (with less political comprehension, information, and voice), not as easy to mobilize for political action compared to urban dwellers (including laborers, civil servants, students, the police, and the army), and less of a physical threat to the state (scattered as they are in remote rural communities). Such factors tend to create a persistent urban bias in the policy-making processes of all developing countries (Lipton 1977). This bias is particularly evident in African governments. They skimp not just on public investments in agricultural development; they also neglect needed investments in rural farm-to-market roads, village water and sewer hookups, rural power and transport, country schools, and rural medical clinics.

Yet even against this more generalized pattern, the neglect of agriculture by governments in Africa stands out. Two-thirds or more of all citizens in Africa depend on farming for income and employment, yet governments continue to devote an average of just 5 percent of their annual budget spending on any kind of agricultural development. In many cases agricultural spending is actually declining in Africa. Whereas the government of Uganda devoted 10 percent of its budget to agriculture in 1980, since 1991 agriculture has not received more than 3 percent of the budget in any year, and in some years the share has been below 2 percent (Oxford Policy Management 2007). African governments are sufficiently ashamed of their performance to at least make regular promises they will do better. At an African Union (AU) meeting in 2003 in Maputo, they pledged to increase their budgetary spending on agriculture to 10 percent within the next five years in support of a new Comprehensive Africa Agriculture Development Programme, known as CAADP (IFPRI 2004). This sounded reassuring, but these same countries had done nothing to fulfill an earlier 1985 pledge to increase agricultural spending up to 20–25 percent of total budgets. At least the new empty pledge was not quite so far-fetched.

Africa needs larger public investments in agriculture and agricultural science more urgently than the rest of the developing world, yet governments in Africa are curiously more neglectful of such investments than governments elsewhere. One important explanation for this dysfunction is Africa's continued dependence on foreign aid in a new international context where so many donors have virtually stopped giving money for production agriculture.

Africa's Distinct Dependence on International Assistance

Dependence on foreign aid has fallen dramatically in recent decades for most governments in Asia and Latin America, but relying on external donors remains an inconvenient fact of life for most governments in sub-Saharan Africa. On a per-capita basis, Africa receives three times as much foreign aid as any of the other developing regions. As a percentage of gross domestic product (GDP) Africa is even more dependent on aid (Devarajan, Rajkumar, and Swaroop 1999). Measures for 1970–1993 constructed by Craig Burnside and David Dollar show that a sampling of twenty-one nations in sub-Saharan Africa were on average more than four times as dependent on aid, relative to GDP, as a sampling of thirty-five developing countries outside of sub-Saharan Africa (Burnside and Dollar 2000). The average sub-Saharan African country now derives roughly 13 percent of its entire GDP from foreign aid, which is five times the foreign-aid dependence experienced by the recovering nations of Western Europe after World War II at the height of the Marshall Plan (Mwenda 2005). African governments need foreign aid simply to service repayment of past debts. Of the forty "heavily indebted poor countries" recently permitted to seek debt relief under a World Bank/IMF initiative, thirty-three were African countries. In other words, thirty-three out of the nearly fifty countries in Africa currently depend on foreign donors simply to service obligations on past debts. New spending for development is seldom undertaken by African governments without some prior assurance of new donor support.

Because African governments are so dependent on foreign assistance and debt relief, and because transfers of foreign assistance into the state budgets of Africa are only partly fungible, spending decisions in Africa tend to be driven either directly or indirectly by donor preferences. When rich donors began cutting their assistance to agriculture in the 1980s, governments in Asia and Latin America were little affected, but aid-dependent governments of Africa were unfortunately dragged along.

Reduced International Assistance to Agriculture

Beginning in the 1980s international donors began cutting back on their assistance to agricultural development. As late as 1980 the U.S. government was still a strong international champion for agricultural development assistance, devoting a full 25 percent of all its official development assistance (ODA) for that purpose. Then the bottom fell out. By 1990 the agricultural share of bilateral ODA from the United States had fallen to just 6 percent of the total. By 2003, to just one percent (Pardey, Alston, and Piggott 2006). Levels of U.S. foreign assistance overall were not falling; in constant dollar terms between 1980 and 2003, total bilateral ODA actually increased by 69 percent. But spending for agriculture simply evaporated.

The downgrading of agricultural development assistance at USAID is now so complete that the word "agriculture" is seldom used anymore inside the agency. Walter Falcon and Rosamund Naylor pointed out that when USAID presented its new sixty-three page five-year joint strategic plan in 2003, the document never directly mentioned agriculture (Falcon and Naylor 2005). The few agricultural programs still being funded are now buried deep inside a hodgepodge category of activities generically labeled "economic growth." USAID still has an agricultural office, but as of 2007 its budget totaled just $27 million, and U.S. development assistance to agriculture from all offices at USAID totaled only $169 million, just a bit more than 1 percent of all USAID spending.

Perversely, agriculture has been falling out of favor inside the U.S. foreign assistance community even as issues of hunger—especially in Africa—have been attracting much more attention. The U.S. government remains prepared to offer ever-larger donations of food aid to hungry people in Africa, even as it has lost interest in helping the African farm sector. Andrew Natsios, a former USAID administrator, openly lamented this reality in 2006; he noted that in the previous year "the U.S. government spent over $1.4 billion on food aid to Africa . . . but only $134 million on agriculture programs to enable Africans to grow their own crops and end recurring food crises" (Natsios 2006a, p. 24). In 2002, just prior to the World Summit on Sustainable Development in Johannesburg, USAID did at last announce a new presidential Initiative to End Hunger in Africa (IEHA), with a goal of cutting hunger on that continent in half by 2015. Yet only part of the modest $200 million in funding assigned to this initiative was new money, and roughly half depended on monetized proceeds from food aid donations.

The share of U.S. ODA going to promote agriculture in developing countries declined after 1980 because Congress, in particular, was losing interest in agricultural development. There were plenty of other seemingly worthy concerns to focus on: women's health, population, the environment, human rights, democratization, microcredit, HIV/AIDS, and postconflict reconstruction. Also, Africa's growing crisis in farm productivity was easy to miss in the mid-1980s, because the well-publicized success of the Green Revolution in Asia and falling world commodity prices left an erroneous impression that food production problems had been solved everywhere. A closer look would have revealed that food and farming circumstances were getting worse in Africa in the 1980s, not better. The lower international commodity prices of that decade reflected growing farm productivity everywhere *except* Africa. Between 1975 and 1985 the number of children in Africa suffering from chronic malnutrition actually increased by 30 percent, from 18.5 million up to 24.1 million (Smith and Haddad 2000).

Agricultural scientists in the United States knew that the agricultural productivity problem had not yet been solved in Africa. They complained about USAID's abandonment of the issue. Dr. Norman Borlaug, winner of the 1970 Nobel Peace Prize for his earlier work on seed improvements for the Green Revolution, testified to the Agriculture Committee of the U.S. House of Representatives in 1994, "Just when the need for greater effort in agricultural development has become so urgent, support to this work is being gutted" (Borlaug 1994, p. 9). But this was the 1990s, and Borlaug's model—using science and technology to boost farm productivity—was badly out of fashion in Congress, so the drop in aid levels continued.

Once the United States began pulling back on bilateral assistance to support agricultural development in the 1980s, it was easy for other rich countries to do the same. Between 1983–84 and 2003–04 overall the share of bilateral aid going to agriculture from the United Kingdom fell from 11.4 percent to 4.1 percent, in France from 8.5 percent to 2.2 percent, and in Germany from 9.1 percent to 2.9 percent (Oxfam 2006). The aggregate value (in constant 1999 U.S. dollars) of all bilateral agricultural ODA to all poor countries from all rich countries fell by 64 percent between 1980 and 2003, from $5.3 billion to just $1.9

billion (Pardey, Alston, and Piggott 2006, calculated from Table 12.4, p. 351). And Africa was asked to absorb these cuts no less than other regions. In real terms, between 1990–92 and 1999–2001 external assistance to agriculture in the developing world as a whole declined by 24 percent, while assistance to farming in the countries with the highest prevalence of undernourishment—mostly African countries—actually declined by 49 percent (Pingali, Stamoulis, and Stringer 2006).

Cutbacks in U.S. foreign aid to agricultural science in Africa were particularly severe. Between the mid-1980s and 2004, annual USAID funding for agricultural R&D at the national level in sub-Saharan Africa fell by roughly three-quarters, down to a negligible $15 million for the entire continent of Africa by 2004. African governments were unable to make up for this tremendous fall, so in about half of the recipient countries agricultural research spending stopped increasing and went into decline (Pardey et al. 2006).

Other donor institutions were following a similar path. The World Bank has always been a leader in championing the importance of agricultural investments in poor countries, including research investments, yet its own lending programs have been heading in the opposite direction ever since the early 1980s. Between 1978 and 1988 the share of lending from the World Bank that went to agricultural development fell from 30 percent to 16 percent. When agricultural advocates criticized this adverse trend, Bank officials agreed it was unfortunate and said it would soon be reversed (Lipton and Paarlberg 1990), but instead the trend worsened. By 2006 only 8 percent of all sectoral lending at the World Bank was going to agriculture (World Bank 2006b). The Bank shed so much of its internal agricultural staff that it lost the technical capacity to design good agricultural loans, which then became another excuse for withdrawing from the sector (Falcon and Naylor 2005). In 2006 the World Bank, which has a staff of roughly 10,000, had only seventeen technical experts assigned to the department that deals with agriculture and rural development in Africa (Dugger 2007). In 2005 Paul Wolfowitz, then World Bank president, admitted in an offhand comment to a business forum, "My institution's largely gotten out of the business of agriculture" (Hitt 2005).

A convergence of political pressures from both left and right beginning in the 1980s led the World Bank to drop lending for agricultural productivity from its portfolio. From the left, activists in rich countries began criticizing the Bank for the environmental damage caused by some past lending projects, especially in the forestry and energy sectors. As a consequence, in 1985 the U.S. Congress passed legislation to make further U.S. financial contributions to the Bank contingent on environmental reforms. The Bank appeased these critics in 1987 by taking four steps: requiring a review of all loans for environmental impact; requiring borrowing countries to prepare comprehensive environmental action plans as a condition to remain eligible for low-interest loans; creating a new environmental department; and creating an inspection panel that gave NGOs a mechanism for participating in the review of projects and blocking those they did not like. These new procedural constraints made new lending for agricultural productivity far more difficult for the Bank, particularly when irrigation projects or agricultural chemical use might be implied.

Powerful pressures simultaneously hit the World Bank in the 1980s from the political right. Reaganite and Thatcherite elements were then holding sway in both the United States and the United Kingdom, and they began criticizing the Bank's traditional emphasis on public-sector investment lending, calling instead for a more market-led approach to development. To appease these forces, the Bank pulled back from public investment lending—including lending for new productivity investments in the agricultural sector—and shifted instead to a pursuit of "structural adjustment" policy reforms meant to get the public sector out of the way of private markets. This state-minimalist approach never delivered the promised growth results in Africa because new public investments were needed there at least as much as market-oriented policy reforms, but when these disappointing results emerged in the 1990s, the response from the political right was to skip over public investment needs again and call this time for institutional reforms in the name of "good governance." To appease criticism from the right, the Bank also tried to adopt a more "corporate" management model that called for loan impacts within a shortened three- to five-year time frame (Falcon and Naylor 2005).

International agricultural science, which does not deliver results within a three- to five-year time frame, was disadvantaged in this new environment. The leading international institution responsible for delivering agricultural science, under the chair of the World Bank, is the Consultative Group on International Agricultural Research (CGIAR), created originally in 1971 when agricultural research was still a highly favored activity among aid donors. This system eventually expanded to include fifteen separate international agricultural research centers, mostly located in the developing world and funded by a collection of bilateral donors, private foundations, and the World Bank. Total annual funding for the system increased tenfold during the 1970s and then doubled once again in real terms in the 1980s, eventually reaching an average annual level of $337 million by the end of that decade (Alston, Dehmer, and Pardey 2006).

Then, as a part of the larger defection of prosperous countries from funding agricultural science, donor support for this CGIAR system faltered. By the 1990s the production success of the Green Revolution in Asia had spawned a view that hunger was now simply a "distribution" problem (an erroneous view for Africa, where most food consumers depended on their own production and where per-capita production was actually falling). Environmentalists, meanwhile, were rejecting the more productive model of the Green Revolution over anxieties about irrigation dams and harmful chemical use. The CGIAR system tried to appease these concerns by shifting its focus away from agricultural productivity research and by creating new research centers to address resource concerns such as forestry and water management, but donor support was still hard to sustain. The United States, once the greatest champion of the CGIAR system, was again the first to pull back. The real dollar value of U.S. contributions to CGIAR fell 47 percent between the early 1980s and the late 1990s. Japan's strong support for the system eventually faltered as well, dropping by 50 percent following that country's own economic crisis later in the 1990s.

The CGIAR response to this funding crisis was to downgrade even more the work of its original crop-production centers. During the 1990s overall, nine of the centers (including all four of the original centers focused on productivity) experienced budget cuts rather than budget expansion. The share of funding that went to enhance agricultural productivity fell from its original level of 74 percent in the mid-1970s to just 34 percent by 2002 (Alston, Dehmer, and Pardey 2006). Expenditures on productivity-enhancing agricultural research declined in real terms by 6.5 percent *annually* between 1992 and 2001 (Falcon and Naylor 2005).

The reductions in U.S. and Japanese support for the CGIAR system had another effect. By 2004 it was the European nations as a group, including the Commission of the EU, that were left providing 41 percent of total funding. This "Europeanization" of the funding base for public international agricultural research would emerge later as a serious barrier to research on genetically engineered crops, once Europe as a whole began rejecting this new technology in the late 1990s.

Hostility to Green-Revolution Farming among Some NGOs

African governments not only suffered a withdrawal of official development assistance to agricultural science after the 1980s; they also encountered an increase in international NGO hostility to science-intensive farming. The domestic social movements that had arisen to criticize industrial agriculture in rich countries in the 1980s could not resist the temptation to warn poor countries—even those starved for science—against modern farming techniques as well. These warnings were easily brushed off by the developing countries of Asia, where farmers had already tasted the benefits of the Green Revolution and where local capacity was high enough to diminish the need for assistance from NGOs. In Africa, however, most local farmers had not yet seen the benefits a technology upgrade might bring, and dependence on international NGOs for service delivery remained high, so campaigns against science-based farming had a significant impact.

Not all international NGOs working in the area of food and agricultural policy are hostile to science. Oxfam International, a strong organization with a clear-eyed focus on hunger and poverty concerns, is not afraid to stress the importance of agricultural productivity in Africa. In 2006, Oxfam openly lamented the fact that "aid for agricultural production in sub-Saharan Africa dropped by 43 percent between 1990–92 and 2000–02" (Oxfam 2006). Bread for the World, a faith-based international NGO focused on hunger, has also embraced the importance of helping farmers in Africa become more productive; this organization launched in 2007 a new campaign titled "Help Farmers. End Hunger." FARM-Africa, with international headquarters in London and offices in Ethiopia, Kenya, Tanzania, and Uganda, argues that greater investments in smallholder agriculture in Africa, including in the "intensification of production," should be the main path out of poverty for millions of poor people (FARM-Africa 2007). Yet many international NGOs do not take poverty and hunger as a starting point. Many are in the business of exporting to Africa visions of environmentalism, anticorporate populism, and organic food purity that exclude science and are a mismatch to Africa's needs.

NGO warnings against agricultural science are consistently broadcast into Africa today through the special agencies of the United Nations system and through global conferences sponsored by the UN, where NGOs have come to enjoy a semiofficial status. The UN system is valued by African governments because it treats each of them as though they were a sovereign power equal to the big countries, and because it permits them—through bloc voting—a rare opportunity to exert genuine international influence. The UN system is also convenient for African states with minimal diplomatic capacity as a venue for maintaining formal ties with other states at an affordable cost. The ability of NGOs to use this UN system to influence African governments was greatly enhanced in 1996 when the UN made formal "consultative status" available to international NGOs. By 2003 a total of 2,350 separate NGOs had gained consultative status within the UN system (Hill 2004). With the presence of so many NGOs, special UN events today—such as "food summits" convened by FAO or "environmental summits" convened by the United Nations Environment Programme (UNEP)—are transformed into settings where international civil-society activists preach their message. In the case of agriculture, the loudest message from the NGO community has been to stay away from modern farm science.

At the 1996 FAO World Food Summit in Rome, 1,200 NGOs from eighty different countries convened their own parallel NGO forum to condemn what they called "industrialized agriculture." The science-based farming model of wealthy countries was depicted at this forum as "destroying traditional farming, poisoning the planet and all living beings." The forum endorsed as its preferred alternative a reliance on traditional technologies, agro-ecological farming, and organic agriculture (NGO Forum 1996). At a follow-up UN food summit in Rome in 2002, a reconvened international NGO forum even blamed the Green Revolution for the rise in world hunger, a ludicrous assertion since the largest rise in hunger had been in Africa, the one continent where the Green Revolution had not yet taken off (NGO/CSO Forum 2002). When NGOs converged at the 2002 UNEP World Summit on Sustainable Development in Johannesburg, they had plenty to say about the need to keep GMOs out of Africa, but almost nothing to say about the need to increase the productivity of Africa's small farmers. Partly as a consequence, the final "Recommendations on Sustainable Development" adopted at the 2002 Johannesburg summit made not a single reference to agriculture (WSSD 2002).

In the international NGO community, the strongest criticism of science-based farming usually comes from European and North American organizations dedicated at home to organic farming and agro-ecology, environmentalism, and anticorporate populism. Many are committed idealists who have formed their strong convictions by studying what they don't like about farming in the industrial north. When they bring these convictions to Africa, where science-intensive farming is not even on the horizon, the needs of poor people are seldom well served. Consider, for example, various NGO campaigns to extend organic farming to Africa.

NGO Advocates of Organic Farming in Africa

In Africa, where farmers struggle with low soil fertility, the last thing they need to be told is to avoid all use of chemical fertilizers, yet this is the inflexible rule organic farming lays down. The best technical solution would be an increased use of *both* chemical fertilizers and organic approaches (such as green manuring and composting) on Africa's small farms, but this is blocked under organic dogma. Nonetheless, a number of European governments—including Germany, Switzerland, Sweden,

Belgium, and the Netherlands—use their foreign assistance programs to promote organic farming in the developing world, including Africa, having been pushed to do so by their own domestic organic farming associations (Parrott and Marsden 2002).

The single most prominent international NGO dedicated to organic promotion in poor countries is the International Federation of Organic Agricultural Movements (IFOAM), an organization headquartered in Bonn with more than 730 member organizations in over 100 different countries, plus an Internet training platform funded by the Dutch government. The mission of IFOAM is not to boost farm productivity but to enlist farmers in the organic movement. In 2006 Mette Meidgaard, IFOAM's vice president, stated to the UN, "The major constraints to achieving universal food security are found in social, economic and political conditions, more than in problems regarding agricultural productive capacity" (IFOAM 2006). The inflexibility of the organic approach obliges IFOAM representatives to argue against any use of nitrogen fertilizer, even in Africa where the depletion of soil nutrients is a severe constraint on production. Africans are told instead to invest their labor in composting animal manure, just like organic growers do back in Germany or the United Kingdom. From within the mindset of IFOAM, Africa's current underutilization of chemical fertilizers is actually seen as a major opportunity. Because so few African farmers are currently using chemical fertilizers, it will be much easier to get them certified as organic. Poor and nonproductive, but certified organic.

Greenpeace International is also a strong promoter of organic farming in poor countries. An Amsterdam-based environmental NGO with a substantial annual budget of €169 million, Greenpeace has national or regional offices in forty-one different countries around the world and claims 2.8 million individual supporters, mostly Europeans. The chief scientist at Greenpeace, Doug Parr (a chemist), argues that the *de facto* organic status of most smallholder farmers in Africa represents a major opportunity to lock in patterns of low chemical use. He believes international activists should help African farmers "develop self-confidence in their traditional knowledge so that they do not immediately switch to chemicals once they can afford them" (Parr 2002, pp. 6–7). Parr admits synthetic chemical fertilizers can bring production gains, but he believes—while giving no evidence—these gains will only be temporary because of the environmental degradation such fertilizers would cause. He manages to ignore the production gains from synthetic nitrogen fertilizers that have been sustained in rich countries for roughly a century now, with no end in sight. For Greenpeace, however, it isn't really about production. Parr dismisses the importance of agricultural production by saying, "There is no direct relationship between the amount of food a country produces and the number of hungry people who live there" (Parr 2002, p. 4). Parr, like so many others, is importing into Africa a half-truth about production and hunger that applies best in Latin America.

Some NGOs even like to promote European-style organic farming in Africa as though it were somehow authentically African. A German organization named Networking for Ecofarming in Africa (NECOFA), which is an arm of the Agriculture Development Centre of the German Foundation for International Development (DSE), has established partner groups in thirteen African countries to warn them that "Western agriculture" cannot provide solutions in Africa where "indigenous farmers' knowledge" should instead be the key. Yet when conducting its training workshops in Africa, this organization likes to teach the chemical-free farming principles developed back in Austria by Rudolf Steiner, the mystic and originator of biodynamic farming. German trainers at one NECOFA session in Kenya in 2005 took the time to introduce local participants to the importance of light rhythms from the planets and to instruct them in developing manure preparations that included essential bits of stinging nettle, chamomile, and cow horn (NECOFA 2005). Such knowledge is neither farmer-derived nor indigenous to Africa, nor is it even knowledge.

Promoters of organic farming in Africa reject agricultural science as misguided because it does not begin with "local knowledge and tradition" (Parrott and Marsden 2002, p. 10). Yet in Africa the traditional local method of restoring soil fertility was to leave a piece of land fallow and uncultivated for up to a dozen years, an option that is no longer available to farmers due to increasing population pressures on the land. Because fallowing time has now been drastically shortened, sometimes to just a year or two, nitrogen is being removed from Africa's soils at an average annual rate of 22–26 kg per hectare per year (Smaling et al.

2006). This severe soil-nutrient depletion—known as "soil mining"—results in crop yields that are not just low but in some countries actually declining (Chrispeels 2004). Annual soil nutrient balances throughout Africa are now negative, causing crop losses every year estimated at between $1 billion and $3 billion. As soil nutrients become exhausted, infertile fields are abandoned to weeds, and new lands, including forest and fragile savanna, must be cleared, leading to still more environmental degradation. Land clearing for agriculture has been estimated as the cause of approximately 70 percent of all deforestation in Africa (WRI 1992).

To increase crop production on the farm and protect land and wildlife habitat off the farm, Africans should be using much more chemical fertilizer rather than less. To reach a goal of 6 percent annual growth in farm production in sub-Saharan Africa by 2015 (the goal recently set by the New Partnership for Africa's Development [NEPAD]), average fertilizer application rates will have to increase sharply from the current level of 9 kg per hectare up to a level five times higher—49 kg per hectare (Camara and Heinemann 2006). Yet organic farming advocates in Africa would apparently prefer to go in the opposite direction, reducing chemical fertilizer use to zero.

At present nearly all certified organic production in Africa is destined for export, intended for consumers in Europe. Specialty crops such as avocados, coconuts, coffee, tea, fruits, and vegetables are grown organically in Africa, often on highly specialized industrial-scale farms near ports or airports to ensure quick transport to supermarkets in Europe (resembling earlier colonial trade patterns). Africa's rural poor gain little from such activities; Africa's smallholder farmers need to productively grow maize, yams, or cowpeas on their own farms, not avocados for somebody else. Organic farming advocates from IFOAM nonetheless like to assert that organic agriculture in developing countries is not a luxury but somehow a precondition for attaining food security.

U.S. NGOs Opposed to Agricultural Science

It is not only European NGOs that have become hostile toward the promotion of science-based farming in the developing world. One of the most visible international NGO campaigns against farm science in Africa has been run by the Institute for Food and Development Policy, a Californian organization better known as Food First. Food First got its start in the 1970s by warning Americans against excessive meat consumption. Its founder was Frances Moore Lappe, a 1966 graduate of a Quaker college in Indiana who worked for a time as a community organizer in Philadelphia before becoming a countercultural food activist. In 1971 Lappe wrote a best-selling book (3 million copies sold) titled *Diet for a Small Planet.* The central argument of this book was that meat consumption in rich countries was using up scarce land resources to grow vast quantities of grain for the feeding of chickens, pigs, and cattle, even as people in poor countries starved; the answer was ecological vegetarianism (Lappe 1971).

This original vision had little to offer Africa. If rich countries stopped eating meat, their land, no longer needed to grow grain for livestock, would not be used to feed poor Africans. Nobody would step forward to pay farmers to plant for that new purpose or pay grain companies to export for that purpose. Grain is not a natural resource; if commercial demand goes away, supply goes away. Moreover, any kind of widespread vegetarianism in Africa itself would be a food-security nightmare. Meat animals in Africa are not a burden on the human food system but frequently the only way to secure adequate human foods from dry grazing lands that are useless for crop production.

Lappe's views quickly evolved, however, into an antitrade, antiscience posture. She wrote another best-selling book in 1977 with Joseph Collins titled *Food First: Beyond the Myth of Scarcity,* in which she rejected international trade as a solution to hunger, championing instead a vision of local food self-sufficiency. She did not, however, advocate self-sufficiency based on modern farming techniques, which she warned would lead to land evictions, debt bondage, and other social outcomes damaging to the rural poor (Lappe and Collins 1977). Lappe and Collins were again likening all of the developing world to the most land-unequal nations of Latin America, and they ignored evidence already becoming available that farming during the Green Revolution had helped the poor in Asia and could help the poor in Africa as well (Thirtle, Lin, and Piesse 2003).

Strong opposition to Green Revolution–style farming in poor countries nonetheless became a central project of Lappe's NGO (named Food

First), created in 1975. Working primarily as an advocacy think tank, Food First identified "intensive, externally dependent models of production" as a cause of deepening poverty and growing hunger around the world. The alternative to be promoted was "food production for domestic and local markets based on peasant and family farmer diversified and agroecologically based production systems" (Food First 2002). The problem with this alternative is that it is a perfect technical description of the nonproductive, science-starved smallholder farming systems that operate in most of rural Africa today. What Food First seems to endorse for the African countryside is little different from the impoverished status quo. Peter Rosset, the current executive director of Food First, dismisses the importance of producing more food in Africa and echoes the Greenpeace line: "Hunger," says Rosset, "is not caused by a shortage of food, and cannot be eliminated by producing more" (Rosset 2000).

Despite the mismatch of Food First's vision to Africa's needs, the organization manages to enjoy enormous credibility among nonspecialists in both the United States and Europe. The *New York Times* has described Food First as one of the nation's "most respected food think tanks," and Lappe herself has received seventeen honorary doctorate degrees from distinguished institutions. In 1987 in Sweden she received the Right Livelihood Award (known as the "alternative Nobel") for her "vision and work healing our planet and uplifting humanity."

Other NGOs from the United States are working to extend to Africans their anxieties about the motives and actions of big agribusiness companies. This makes populist sense in America, but it is a dubious preoccupation to project into the impoverished reaches of rural Africa, where salesmen from big multinational seed and chemical companies are almost nowhere to be found. The Institute for Agriculture and Trade Policy (IATP), based in Minnesota, was founded in 1986 as a populist advocacy organization warning against the dangers of corporate concentration and free trade in American agriculture, but IATP has not been able to resist the temptation to take this same message to farmers in poor countries, where it warns that the Green Revolution model is a dangerous "technical fix" that will only bring profits to the corporate sector (Murphy 2006).

IATP may be unaware that the original Green Revolution was brought to Asia not by private corporations but instead by the Rockefeller and Ford Foundations and a host of public-sector donors, researchers, and extension workers—and even some NGOs. Having seen agricultural science and agribusiness working hand-in-glove back in Minnesota, IATP assumes these co-conspirators are now posing a threat to Africa as well. IATP's fears of corporate domination are wildly out of place in Africa. Recall that only 1.7 percent all current agricultural R&D spending in sub-Saharan Africa comes from the private sector. Africa is such an unattractive place for private companies to work (due to governmental taxes and corruption, weak property protection, unreliable contract enforcement, poor infrastructure, and physical insecurity) that most multinational agribusiness firms make few investments of any kind there. The real danger in Africa is a continuing corporate disinterest in local farm-technology needs, not corporate domination.

Uncertain Support from Philanthropic Foundations

Independently endowed philanthropic foundations—such as the Rockefeller and Ford Foundations—played an essential role in launching Asia's Green Revolution almost half a century ago. Times have changed. The Ford Foundation moved away from promoting scientific innovations in agriculture in poor countries not long after the Green Revolution succeeded, and the Rockefeller Foundation may now be following suit.

The Ford Foundation is a private, nonprofit, grant-making institution with more than $10 billion in assets, a New York headquarters, and three African regional offices in Lagos, Nairobi, and Johannesburg. The foundation's priorities once included farm science to increase productivity, but no longer; its few remaining agricultural grants today promote "alternative agriculture," and one of its favorite grantees is IATP. Ford now refers to the original Green Revolution it helped launch in Asia as only a limited success, and one that brought unintended negative environmental impacts linked to "the large amounts of fertilizer and pesticides needed to grow the high-yielding crop varieties" (Ford Foundation 2006, p. 38). One recent Ford grantee in India is a local organization devoted to convincing large landowners in Uttar Pradesh to convert to organic farming.

Whereas the Ford Foundation abandoned funding technological solutions to low farm productivity quite early, the smaller ($3.4 billion endowment) Rockefeller Foundation remained faithful to this vision right up through 2005. Rockefeller was led between 1998 and 2005 by Gordon Conway, an agricultural ecologist who promoted what he called a "doubly green revolution," one that would be based on the best agricultural science supplemented by new attention to environmental effects. During Conway's term as president the Rockefeller Foundation spent a total of nearly $150 million promoting agricultural development specifically in Africa, bringing a significant scientific focus to the problem through the work of twenty-five different plant-breeding teams producing food-crop varieties with higher yields and greater disease resistance (Rodin 2006).

When Conway left in 2005, the Rockefeller Foundation began looking for a way to hand off its traditional lead role in agriculture and found that way in 2006 when the Bill and Melinda Gates Foundation, with an endowment ten times as large, made an important move into the agricultural development field. In September 2006, the Rockefeller Foundation announced it was entering into a $150 million joint venture with the Gates Foundation ($50 million from the Rockefeller Foundation, $100 million from the Gates Foundation) that would be called an Alliance for a Green Revolution in Africa (AGRA). Kofi Annan agreed in June 2007 to serve as AGRA's first chairman. Existing Rockefeller programs would provide the basis of the effort, but Gates Foundation money would provide a substantial boost to the scale of activities. The main thrust would be an across-the-board effort to improve the variety and availability of seeds capable of producing higher yields in the harsh conditions of sub-Saharan Africa. Bill Gates himself explained the importance of introducing African farmers to modern agricultural science:

> In Africa today, the great majority of poor people, many of them women with young children, depend on agriculture for food and income and remain impoverished and even go hungry. Yet, Melinda and I also have seen reason for hope—African plant scientists developing higher-yielding crops, African entrepreneurs starting seed companies to reach small farmers, and agrodealers reaching more and more small

> farmers with improved farm inputs and farm management practices. These strategies have the potential to transform the lives and health of millions of families. Working together with African leaders and the Rockefeller Foundation, we are embarking on a long-term effort focused on agricultural productivity, which will build on and extend this important work. (Bill and Melinda Gates Foundation 2006)

The unmatched philanthropic resources of the Bill and Melinda Gates Foundation may yet be able to turn the tide of fashion back toward supporting pro-poor agricultural science in Africa. Yet from the NGO community the first response to this new Gates initiative was predictably hostile. Peter Rosset at Food First lampooned the Gates' "naiveté about the causes of hunger" and warned that the most likely result of the new initiative would be "higher profits for the seed and fertilizer industries, negligible impacts on total food production and worsening exclusion and marginalization in the countryside" (Rosset 2006). GRAIN, an NGO concerned with agroecology and genetic diversity headquartered in Barcelona, scolded Bill and Melinda Gates for thinking that increased fertilizer use might be of any use to Africans and cited a letter sent earlier by more than 600 NGOs to the Director General of FAO which said, "If we have learned anything from the failures of the Green Revolution, it is that technological 'advances' in crop genetics for seeds that respond to external inputs go hand in hand with increased socioeconomic polarization, rural and urban impoverishment, and greater food insecurity" (GRAIN 2006).

Caring about Africa, but Not about Agriculture

Agricultural advocates who hope the international tide can be turned back in favor of science-based farming know there is at present no lack of concern in wealthy countries for Africa itself, only for the productivity of African agriculture. Ordinary citizens in prosperous countries have become concerned with Africa as never before, and their governments have noticed. Largely in response to popular demands mobilized by citizen groups and celebrities, and fanned by the media, the leaders of the G8 countries, meeting in Scotland in July 2005, agreed to double the annual foreign assistance they would give to Africa within the next five years, up to a level of $50 billion by 2010.

A prominent community of international celebrities concerned about the welfare of Africa has emerged to push this assistance agenda forward. To promote the G8 outcome, the Irish rock star Bob Geldof had organized a series of free "Live 8" concerts watched by an estimated 3 billion people around the world. Geldof had tireless promotional help from his remarkable countryman Bono, lead singer for U2, who has also pushed for help to Africa through an organization he co-founded named DATA (Debt AIDS Trade Africa), chaired by Bobby Shriver, a nephew of former president John F. Kennedy. Ordinary citizens have responded. An estimated 1.4 million Americans were among those who joined the campaign to force the final G8 result in Scotland, taking the lead from their church leaders on Sunday morning as well as from the Hollywood stars now flocking to celebrity-dotted Africa fundraisers on Saturday night (ENS 2005).

In 2006 the *New York Times* made a shallow effort to debunk this new African interest among the rich and privileged, labeling it a self-serving form of "misery chic." Self-promotion clearly played a role: George Clooney campaigning to end the killings in Darfur; Madonna helping to finance orphanages in Malawi (adopting an adorable child who turned out to have a living father); Angelina Jolie going to Kenya to film an MTV special; Gwyneth Paltrow's face appearing in a full-page advertisement over the words, "I Am African" (in the fine print we see it is an AIDS appeal) (Rice 2006). Numerous other stars sought publicity by making their own personal pilgrimages to Africa: Drew Barrymore, Matt Damon, Mia Farrow, Ashley Judd, and Brad Pitt. Oprah Winfrey at one point did everything short of actually moving to Africa. Yet this sudden eagerness by celebrities to be seen in Africa was itself a proof that growing numbers of citizens in rich countries had become committed to the cause.

Some of the efforts now being made by ordinary citizens and governments in rich countries to reach out to Africans in need are simply without precedent. Consider the resources recently made available to fight malaria, a disease that kills 800,000 children in Africa every year. In 2005 the World Bank committed $350 million to combat malaria in ten countries in West Africa. Then the Bill and Melinda Gates Foundation promised to spend $766 million to fight this disease. Then President George W. Bush trumped them all with a pledged $1.2 billion for the five-year President's Malaria Initiative (Dugger 2006). Ordinary citizens have also organized to do their part. Volunteer youth groups in the United States with names like "Veto the 'Squito" and "Nothing but Nets" invest street-level time and energy soliciting and aggregating individual $10 donations to help pay for malaria drugs and mosquito bed nets. This mobilization of resources by so many ordinarily self-absorbed people in rich countries has enabled aid-dependent Africa to do a much better job of meeting urgent public health needs. As recently as the 1990s, governments in sub-Saharan Africa typically spent less than 3 percent of their budget on public health. By 2003 Tanzania was spending 13 percent; Namibia and Zambia, 12 percent; and Uganda, 11 percent (Garrett 2007).

Given the obvious desire to help Africa, and given the clear evidence that external assistance can generate a stronger local policy response, it is all the more discouraging to find official development assistance to African agriculture still in such a collapse. It seems that rich countries find it easier to respond to the health dimension of Africa's current crisis because human health remains a central concern back home. The agricultural dimension of Africa's crisis has regrettably become more difficult to comprehend in wealthy countries. Having evolved so far beyond any further need for agricultural productivity at home in Europe and North America, there is little natural inclination to support agricultural research in Africa. Saturated with too much agricultural science at home, Americans and Europeans have developed a tin ear for the needs of science-starved farmers in Africa.

4

Keeping Genetically Engineered Crops Out of Africa

As with agricultural science overall, so with GMOs in particular. Citizens and government officials in prosperous countries do not value agricultural GMOs at home, so they presume to draw a similar conclusion for Africa. They either make little effort to bring this new science to Africa, or they work actively to keep it away. As a consequence, this new agricultural science is not yet available to most farmers in Africa. As of 2007 GMOs were being grown commercially in only one country on the continent—the Republic of South Africa. None of the other fifty-three countries of Africa had yet made it legal for farmers to plant any GM crops.

In the previous chapter I observed that when wealthy countries began withdrawing their international assistance to agricultural science, the aid-dependent governments of Africa took the cue and followed the same path. In this chapter I find multiple explanations for Africa's decision to keep modern agricultural biotechnology at a distance, tracing the international influence primarily to Europe. The governments and citizens of Europe continue to exercise considerable postcolonial influence in Africa through a range of mechanisms that include foreign financial and technical assistance, international organization activities, NGO advocacy campaigns, and international commodity markets. Through each of these channels today Europe is telling governments in Africa that it would be best to stay away from agricultural GMOs, and African governments have responded accordingly.

How can we be certain that Africa's resistance to GM foods and agricultural crops is a postcolonial effect introduced from the outside, rather than a preference that is authentically African and home grown? One method is to compare the extremely stringent regulatory standards governments in Africa are now adopting for agricultural GMOs to the far less stringent standards they tend to adopt toward virtually every other food and farming technology. Abnormally stringent standards for GMOs can only be understood as an exotic import into Africa from the outside, in this case from the leading international proponents of stringent regulation, the societies of Western Europe.

In this chapter I begin by examining the uneven uptake of GM crops in the developing world, including Africa, and review several possible explanations for the patterns that emerge. A range of misleading explanations for Africa's failure to take up this technology is considered, but ultimately rejected in favor of an explanation based on postcolonial influence from Europe. The channels through which this external influence is exercised will then be examined in some detail.

The Uneven Uptake of GM Crops in the Developing World

Advocates for genetically engineered agricultural crops like to brag about how quickly the technology has spread in the dozen years since the first varieties were given their initial approval for planting and import by national regulators in North America and Europe in 1995 and 1996. Since 1995, the global area planted to GM crops has expanded at double-digit rates every year, reaching 102 million hectares by 2006, or 252 million acres. Yet this uptake of GM crops has still been limited to only a few crops and it has been highly uneven country by country. As of 2006, only twenty-two countries around the world had any significant commercial plantings of GM crops, and roughly 90 percent of global GM crop area was still confined to just four countries, all in the Western Hemisphere: the United States, Argentina, Brazil, and Canada (James 2006). The United States by itself made up a lopsided 53 percent of the total world crop area planted to GMOs. In addition, nearly all GM crops grown so far have been intended for use either as animal feed or for industrial purposes, rather than for direct human consumption as staple foods. The most frequently planted GM crops so far have been soybeans, yellow maize, canola, and cotton; the first three are heavily used for animal feed or cooking oil, and the fourth is an industrial crop. Soybeans alone make up 60 percent of all GM-cropped area worldwide; maize, 24 percent; cotton, 11 percent; and canola, 5 percent. The only country that is thus far growing a GM variety of a food staple crop for human consumption is the Republic of South Africa, which first approved the production of GM white maize in 2002. Only 8 percent of all GM cropland globally is planted to food staple crops.

Less than one percent of the total land area planted to all crops in the developing regions of Asia, the Middle East, and Africa is planted to any kind of GM crop, even animal feeds. As of 2006, India was planting 3.8 million hectares of insect-resistant GM cotton but no food or feed crops. China was planting 3.5 million hectares of the same kind of GM cotton but nothing else. Only the Philippines was planting 0.2 hectares of GM yellow maize, Iran just a token area of GM rice, and South Africa 1.4 million hectares of yellow maize, white maize, soybeans, and cotton.

This limited uptake of GM crops in the developing world has been most surprising because of its primary cause: a resistance to the technology by political elites and governments in the receiving countries. Most GM crop advocates in the mid-1990s worried that uptake would be hindered by weak incentives. Private companies would have to develop GM crops appropriate for the poor in tropical environments, or high seed prices and intellectual property constraints would be placed on the technology, or crop scientists in poor countries would have difficulty adapting the technology to local needs. The original advocates never dreamed the technology would go unused because governments in poor countries would brand it as unsafe and refuse to approve its use.

No Shortage of Useable Technologies

Most GM crops now on the market were developed by the private sector with the needs of commercial farmers in prosperous, temperate-zone countries uppermost in mind. This has been one sad consequence of the weak funding recently provided to public-sector agricultural research institutions, especially in poor countries. Yet the original fear that private companies might not produce any genetic technologies of value to small farmers in the developing world was invalidated when poor farmers in China began planting GM cotton in 1997. This was an insect-resistant variety of cotton, called *Bt* cotton because it has been engineered to carry a gene from the soil bacterium *Bt* that expresses a protein certain caterpillars cannot digest. Farmers who plant *Bt* cotton do not have to spray as much chemical insecticide on their fields to control insect pests, which allows them to save money and reduce their occupational exposure to poisons. Varieties of *Bt* cotton are now being planted successfully not only by most U.S. commercial cotton farmers, but also by cotton growers in China, India, and South Africa.

After China first made it legal for farmers to plant *Bt* cotton in 1997, survey data collected in Hebei and Shandong provinces showed spectacular average gains in farm income of $357 per hectare in 1999, $650 per hectare in 2000, and $502 per hectare in 2001, along with positive environmental impacts and fewer family health problems associated with pesticides (Huang et al. 2002). Farmers in India were first given official permission to use *Bt* cotton in 2002, and plantings by both small and large growers increased rapidly in the years that followed, rising 192 percent between 2005 and 2006 alone. In March 2004, one commissioned report found that switching to *Bt* cotton in India increased net profits per hectare for farmers by roughly 78 percent (Krishnakumar 2004).

In the Republic of South Africa GM cotton was first approved for release to farmers in 1997, and within five years roughly 45 percent of all cotton lands in the country were planted to GM seeds. Small as well as large farmers shared in the benefit. A 2002 survey of nonirrigated small farmers in Makhathini Flats in KwaZulu-Natal revealed that switching to *Bt* cotton had sharply reduced the cost of protecting against insects, gaining farmers a net income advantage of $50 per hectare (ISAAA 2002). Currently in South Africa roughly 90 percent of all cotton land is planted to GM varieties, yet on the rest of the continent this successful technology has gone unused because the commercial planting of *Bt* cotton has not yet been approved by any other African government.

As with *Bt* cotton, so with *Bt* maize, which also helps farmers protect against insects with fewer insecticide sprays. Maize is an important staple food crop in eastern and southern Africa. In the Republic of South Africa, where *Bt* maize has been approved, farmers have used it successfully as an important new tool for protecting against damage from stalk borer pests, with net income gains under irrigated conditions equal to $36 per hectare and a net income gain under dryland conditions of $27 per hectare (Kirsten and Gouse 2003). Planting and consumption of *Bt* white maize, now the world's leading example of a GM variety of a staple food crop, has also spread rapidly in the Republic of South Africa. Unfortunately, no other country in Africa has yet approved the commercial planting of any kind of *Bt* maize. In the Philippines, the only other developing country outside of the Western Hemisphere to have approved *Bt* maize, the story is the same. Small farmers switching to this GM variety have recently experienced a 37 percent increase in yields, a 60 percent reduction in insecticide costs, and an 88 percent increase in profitability per hectare (Yorobe and Quicoy 2004). Such examples confirm that the limiting factor on GM crop uptake in the developing countries of Africa and Asia has not been the small menu of technologies on the shelf, but rather the reluctance of most national governments in these regions to approve the planting of any GM crops.

Patent Constraints

A second misleading explanation for the slow uptake of GM crops in Africa and Asia is the supposed reluctance of private biotechnology firms to share their proprietary technologies. It is actually in the poorest countries, where commercial markets for seeds are so small, that the big companies are most often willing to license their proprietary technologies on a royalty-free basis for local use, since they have no significant sales to lose. Monsanto did this in 1991 for a disease-resistant GM sweet potato in Kenya; Bayer AG, Monsanto Co., Orynova BV, and Zeneca Mogen BV did this for a high beta-carotene Golden Rice technology in 2000; and DuPont/Pioneer did this for nutritionally enhanced sorghum in Africa in 2005.

Even in some of the larger and more prosperous developing countries—such as China, India, and Argentina—the biotechnology companies have been more than willing to share intellectual property through joint venture or licensing agreements with local firms, despite the significant risk of local piracy. Nongovernmental organizations have tried to spread the myth that GM seeds are sterile—because they contain an alleged "terminator gene"—but the truth is that GM seeds are just as easy to replicate as non-GM seeds. "Terminator technologies" exist on paper, but they have never been incorporated into any of the GMOs currently on the market, and biotechnology companies have even pledged not to use them. In fact, local piracy is all too easy. In China at one point in the late 1990s, up to half the spread of GM cotton took the form of farmers replanting their own saved Monsanto seeds rather than by purchasing seed anew. Some farmers also bought pirated Monsanto seed at a discount from Chinese merchants, who used copied versions of Monsanto's distinctive logos, boxes, and seed bags. So it is seldom a lack of access to patented traits that slows or blocks the uptake of highly useful GM crop technologies in the developing world.

Scientific Capacity in Africa

A slightly more plausible explanation for the slow uptake of GM crops in Africa has been the weak scientific capacity of these countries, a condition worsened by the post-1980s withdrawal of external assistance to local agricultural R&D discussed in the previous chapter. Local capacity is especially important for agricultural science, since all crop technologies must be custom-tailored to local growing conditions, including soil, moisture, temperature, altitude, topography, length of season, and disease or infestation constraints (Ruttan 2001). The more advanced developing countries such as South Africa, India, Brazil, and China have now developed a strong capacity to perform the adaptive research steps necessary, and even to develop new crop technologies from scratch (China was the first country in the world to develop hybrid rice). In Africa, unfortunately, these important crop-science capacities are often still missing (Evenson 1996). Africa's crop-science capacity in some cases has actually declined as agricultural universities in Africa and national agricultural research systems have suffered more than two decades of weak international support. Between 1971 and 2000 as foreign assistance evaporated, average spending per agricultural scientist in Africa actually dropped by 50 percent (Beintema and Stads 2004).

Creating GM crops from scratch is of course scientifically demanding, yet African scientists do not have to start from scratch. It is relatively easy using conventional breeding methods to transfer GM traits into local crops from related crops that have already been transformed elsewhere. This may still require a team of trained breeders with access to high-quality local germplasm, some lab and greenhouse facilities, and a functioning system of research stations with field trial capacity, but the missing pieces might readily be provided by international donors or by the CGIAR system, which has a strong conventional breeding capacity and high-quality germplasm collections.

Permission to work with patented traits from GMOs must of course be secured from the private companies—or the universities—that claim the patent rights. As a consequence, the favored institutional approach for introducing GMOs to Africa so far has been the *ad hoc* creation of multiple public-private partnerships (PPP) that bring together the science and funding strength of private foundations and CGIAR centers, the local knowledge and credibility of national research systems, and the even greater science capacity and intellectual property claims of private global firms.

So far PPPs of this kind have been created to develop several kinds of GM rice for poor countries, plus GM potato, sweet potato, cassava, and sorghum specifically for Africa (Spielman, Cohen, and Zambrano 2006). By one count, the CGIAR system is currently linked into fourteen such partnerships (World Bank 2008). Yet to date none of the GMOs developed in this manner has been approved for commercial use by Africa's national regulators. In some cases, regulators have not yet even approved field trials for these locally developed GM varieties (Cohen 2005). Not all of these projects produced successful local cultivars, yet we cannot blame the blocked uptake of GM crops in Africa on a failure to mobilize scientific resources. Far more important has been the reluctance of African governments to go forward with regulatory approvals.

Regulations that Are Either Missing or Excessive

The only country in sub-Saharan Africa to have given official approval for the commercial planting of any GM crops so far is the Republic of

South Africa, which took this step in 1997. Farmers in South Africa have had a good experience with GMOs so far and no safety mishaps have been encountered, yet the other governments of Africa have not followed South Africa's lead. Superficially it can appear that these other states are simply waiting to have all of the right laws on the books. One inventory taken in 2006 showed that only two countries in sub-Saharan Africa—South Africa and Zimbabwe—had in place both a biosafety statute and the full regulatory regime needed to implement the statute. Several of the rest, like Kenya, had regulations but no law, whereas others, like Mauritius, had a law but no regulations. Some, such as Namibia, had a national biosafety policy, but neither a law nor any regulations. The majority had no biosafety law or regulations beyond draft form, or even an official policy (ABSF 2006).

In fact, the problem isn't so much a lack of legal authority as it is an incipient hostility to the technology, and an embrace of regulatory systems and attitudes so precautionary as to keep GMOs out of the fields and off the market entirely. Ingo Potrykus is a Swiss plant scientist who led the development in 2000 of a GM variety of rice high in beta-carotene (known as Golden Rice), a product whose development was slowed by the reluctance of national regulators in Asia to approve field trials. When Potrykus saw in 2007 that the World Bank was planning to blame the slow uptake of GM crops in poor countries on "weak regulatory capacity," he sent off the following tart rebuttal: "Where I cannot agree at all, is the notion that 'weak regulatory capacity' is a major cause. It is true that regulatory authorities may have a negative impact, however, not because of weak capacity, but because of the principle of 'extreme precautionary regulation' which frightens any person involved, to make a mistake, leading to the psychological situation that it is better not to take a decision instead of one which could be criticized by the GMO opposition" (Potrykus 2007). Particularly in Africa, the embrace of "extreme precaution" when regulating GM crops is what has kept this technology away from farmers so far.

In its regulatory approach to GMOs, Africa has been following Europe rather than the United States. The United States decided to regulate GMOs for food safety and biosafety using the same statutes in place for conventional crops and foods, but Europe took a different path by creating a legally separate and more demanding system for GM crop regulation (Jasanoff 2005). In part this reflected a more general European preference to regulate production processes as well as products, but it also reflected new moves in Europe toward embracing what is called the "precautionary principle," an approach that allows governments to keep new technologies off the market even if positive evidence of a risk has not been found. Scientific "uncertainty" about the effect of a technology is all that is required to trigger formal disapproval. In practice, resolving every possible uncertainty about a technology's effects is an endless and impossible task; if the precautionary principle were consistently applied (and it never is), it would advise us to "never do anything for the first time." I indicated in Chapter 1 that Europe does not embrace the precautionary principle consistently across all technologies, since medical GMOs are given swift approval even when considerable scientific uncertainty remains as to their long-term or side effects. Only when a technology is not needed by Europeans—such as GM crops—is it likely to get the full precautionary treatment. If the anticipated local benefit is not compelling, a new technology can be slowed down in Europe by what Lawrence Kogan has called an "administratively created presumption of possible harm" (Kogan 2005, p. 5).

Europe's precautionary principle had honorable origins. It first emerged in the context of a serious and well-documented environmental harm in Germany known as forest death. The German government responded with a 1974 clean air act that allowed action to be taken against potentially damaging chemicals even in the absence of scientific certainty regarding their contribution to the harm (Gee, ed. 2002). In 1984 this same principle was then embraced for managing ocean pollution in the North Sea, another documented harm. In 1992, however, the precautionary principle came to be embraced in a more abstract manner in Europe's Maastricht Treaty, and later that same year Europe made it an official part of the Agenda 21 agreement at the UN Conference on Environment and Development (UNCED) in Brazil. It shifted from justifying technological precaution in the face of a documented harm to justifying technological prohibition simply under any uncertainty, without evidence of risk and even without any evidence of documented harm. The European Union now is trying, against U.S. objection, to insert this precautionary principle into the making of food-safety standards by the Codex Alimentarius Commission in Rome, which would automatically introduce it as well into the dispute settlement mechanisms of the World Trade Organization (WTO).

The precautionary principle is only part of the regulatory problem in Europe. The other part is Europe's decision to create new, separate, and more demanding laws for the conduct of risk assessments for GMOs. Under Europe's approach, formal approval for the use of a GMO as food or feed, or for the deliberate release of a GMO into the environment (e.g., planting a GM seed), must now come from an elaborate hierarchy of EU institutions, including the European Food Safety Authority (EFSA), the EU Commission's Standing Committee on Food Chain and Animal Health, and in cases of disagreement, the EU Council, with consultation of the European parliament. Small wonder that so few approvals have recently been given. Approvals are nominally based on technical reviews by specialists (e.g., agronomists, molecular biologists, entomologists, ecologists, and toxicologists) who examine dossiers of test results submitted (and paid for) by applicants (the companies who have developed the technology), but even when no evidence of risk is found, additional tests will frequently be required. For example, before EFSA gave its approval to a new GM maize variety with resistance to rootworms, it insisted on seeing the results of an additional ninety-day rat-feeding study carried out by an independent toxicology facility. These results then had to be reviewed by toxicology experts at this facility, experts at EFSA, and independent national toxicology experts from New Zealand, Italy, Germany, and England. New non-GM maize varieties going onto the market in Europe undergo none of this special scrutiny.

This European approach to the regulation of GM foods and crops—new and separate laws, new or separate institutions, and an application of the precautionary principle—is so demanding and prone to paralysis that it has barely been functioning even in Europe, as was pointed out by the Dispute Settlement Body of the WTO in 2006, in response to trade blockage complaints from the United States. Whenever developing countries have tried to follow the European approach, regulatory approvals have slowed to a crawl. In India the approval of GM cotton was delayed for several years by a demand from the national biosafety committee for an additional goat-feeding study (for the cotton seeds), a cow-feeding study, a water buffalo–feeding study, a pollen-flow study, a soil microflora study, and poultry- and fish-feeding studies (Pray, Bengali, and Ramaswami 2004). It was only after farmers began planting pirated GM cotton seeds illegally—with dramatic success in protecting against infestations of bollworms—that India's precautionary regulators were finally shamed into giving an official approval. They have so far approved nothing else, despite having found no evidence of actual risk. In Brazil, likewise, an official approval for GM soybeans did not come until after farmers had been growing the crop successfully but illegally for a number of years.

Now being introduced into Africa, the European-style approval system for GMOs is becoming a recipe for paralysis there as well. The European system first requires that a separate law be enacted, which leaves many African governments stuck at the stage of trying to get a comprehensive new biosafety bill through their parliaments. Others find themselves unable to assemble a competently staffed and funded national biosafety committee composed of trained specialists in all the required disciplines with adequate budget for travel, technical support, Internet access, and database management. Where such committees have been formed, they have behaved so far just as Potrykus describes: fearful of being criticized for making a mistake, when presented with a formal application, they stall, ask for more tests, or make no decision at all.

Seeing such European-style approval systems going into place in Africa, technology developers quickly become discouraged. Why try to bring a GM crop into Africa if getting over the regulatory hurdles is going to be such a slow, costly, and uncertain process? Scientists in Africa are hurt most of all by these high regulatory hurdles, since they can least afford the additional expenses implied. Doing all the extra tests necessary to generate the dossier of evidence that precautionary biosafety committees will demand is discouraging enough to big international seed and biotechnology companies with deep pockets; it is completely unaffordable to financially stressed national researchers in Africa.

Developing countries that go down the European regulatory path must now also anticipate burdensome demands in the marketing of any GM products. Since April 2004, the European Union has been setting in place regulations governing the labeling and tracing of all GM foods in the marketplace, *after* regulators have approved them as safe for consumers. All GM products must now carry an identifying label announcing that they contain or were produced from "genetically modified organisms," and all operators in the food chain handling these approved GMOs must maintain for at least five years an audit trail showing where each GM product came from and to whom it was sold, purportedly to facilitate an emergency removal from the market of an approved GMO should it later be found unsafe. These new regulations apply to animal feeds as well as foods, to processed foods even when the GM content is no longer detectable, and to foods with as little as nine-tenths of one percent GM content. These regulations have been affordable in Europe since 2004 only because most GM foods had long since been removed from the marketplace voluntarily. States that copy the European approach to labeling and tracing will also be making a *de facto* decision to keep GM foods and crops out of their own market.

Despite all the burdens, costs, and dysfunctions of the European regulatory approach to GM foods and crops, it is this approach that nearly all governments in Africa—all except the Republic of South Africa—have been moving to adopt. This embrace of a European-style regulatory approach toward GMOs, more than any other factor, is blocking uptake of the technology.

Earmarks of External Influence

The stifling regulation of agricultural GMOs in Africa would be easier to accept if it were genuinely home-grown and authentically African. Unfortunately, Africa's extreme precaution toward GMOs has all the earmarks of a policy preference projected into the continent from Europe. Left to their own devices, African governments seldom adopt extreme precaution in either their food safety or their environmental safety policies, and for good reason. They know importing all of the demanding regulatory procedures used by wealthy countries would be too costly relative to the gains, relative to their own regulatory capabilities, and relative to their most urgent development needs.

When it comes to food safety, African governments usually know better than to impose highly precautionary standards. They often have their hands full simply making sure food is sufficiently available to urban populations at an affordable price. It would be nice to guarantee against all hypothetical risks, but in Africa there are still too many known food-safety risks that deserve higher priority. In Africa's open-air markets, foods are often sold uninspected, unpackaged, unlabeled, unrefrigerated, unpasteurized, and unwashed, and the consequences are often fatal. Every year, 700,000 Africans die of food- and waterborne diseases such as cholera, salmonellosis, entero-haemorrhagic *Escherichia coli* (EHEC), hepatitis A, and acute aflatoxicosis (Foodnavigator 2005). It is these real and present consumer food dangers that African governments worry about first, and when they do so they know better than to take the precautionary approach of shutting down all open-air markets. Only when it comes to regulating the safety of GM foods—which have not yet done any documented harm—do African governments curiously take the highly precautionary European approach.

Likewise, in the area of environmental policy, because African countries have not yet achieved material affluence, they ordinarily attach lower priority to the "luxury good" of environmental protection. In one 2006 ranking of 133 different countries according to environmental outcomes, seven of the lowest-ranking ten countries were in Africa, and only one country from all of Africa ranked among the top seventy (Gabon ranked forty-sixth). African governments naturally put economic development first, even when that means doing a below-average job of protecting the environment (Yale Center for ELP 2006). For governments in Africa to reject GMOs on grounds of environmental safety, even when these crops have done no documented damage anywhere, is thus suspiciously out of character. All the more so since these governments so often ignore the real damage done to rural biosafety by conventional farming practices. The continued expansion of conventional cultivation systems in Africa has depleted soil nutrients, forcing farmers to cultivate new lands, which causes deforestation and habitat loss. The plowing of dry grazing lands has caused massive soil erosion, watershed destruction, siltation of surface water, and death of aquatic life. The continued expansion of livestock grazing on dry lands leads to desertification. These real threats to rural biosafety are routinely (if unwisely) ignored by most African governments. So it is suspiciously out of character for them to impose highly precautionary biosafety standards for GMOs, and for GMOs only.

Telling markers of external influence—specifically European influence—can be glimpsed in some of the details surrounding Zambia's decision in 2002 to begin rejecting GM food aid in the midst of an acute drought emergency. When Zambia's minister of information officially announced the ban, he invoked Europe's precautionary principle (Nicholson 2006, p. 8). When officials in Zambia were then asked why they were worried about GMOs, they pointed to the precautionary position taken originally in 1999 by the British Medical Association. "In Zambia, they are always citing the BMA as the reason [for the decision]," said Dr. Luke Mumba, a senior molecular biologist in Lusaka (NewScientist 2003). And when Zambian scientists later endorsed keeping the ban in place, they referred not only to the precautionary principle and the views of the BMA, but also to conversations they had had with Dr. Terje Traavik, scientific director of the Norwegian Institute of Gene Ecology, who had warned them about "a long list of possible, theoretical risk factors" associated with GMOs (Government of Zambia 2002). The closer one looks at African decision making on the GMO issue, the more one finds such earmarks of external influence, especially from Europe.

This external influence is exercised in Africa through at least four parallel reinforcing channels: foreign assistance, the UN system, international commodity markets, and international NGOs.

Foreign Assistance

Political leaders in Africa depend heavily on foreign assistance, and fully half of all the foreign aid flowing into Africa comes from the European Union or from its member states. Official development assistance from Europe to sub-Saharan Africa is nearly three times as great as assistance from the United States ($11.0 billion from the EU countries together in 2004, compared to $4.1 billion from the United States). It is thus not surprising to see governments in Africa paying greater attention to the policy preferences of European donors on GMO issues than to preferences of the U.S. government.

Although Europe provides generous assistance to Africa, little now goes for agricultural development. In 2005 the European Council laid out its assistance goals for Africa, beginning with peace and security, then moving on to human rights and governance, then to investments in infrastructure, harmonized legal frameworks, environmental protection, education, health, women, and migration. No reference was made anywhere to agriculture. One sentence was devoted to improved food security, yet this was to be achieved through improved "safety-nets" for Africans who were dependent on food aid, rather than through increased food production. What little development assistance Europe does give to African agriculture does not support GMOs. The culture against GMOs among European donors is so strong that it can even compel those who were once supporters of the technology to change their position. Gordon Conway, who gave significant support to GM crop development in Africa as well as Asia during his time as president of the Rockefeller Foundation through 2005, had to curb his enthusiasm for the technology after he became chief science advisor to the U.K.'s Department for International Development (DFID). Conway could now publicly endorse the use of DFID money to promote tissue culture and marker-assisted breeding, but not development of GMOs.

Instead of supporting GMOs, European donor agencies like to support NGOs campaigning against the technology. For example, the Swiss Agency for Development and Cooperation (SDC) and the Netherlands Ministry of Foreign Affairs (DGIS) support an influential project that puts GMOs "on trial" in developing countries in front of locally assembled "citizen juries." These show trials are facilitated by a Swiss biosafety institute named RIBios and the International Institute for Environment and Development, IIED, based in London. The local citizens spend several days hearing testimony from expert witnesses, including anti-GMO activists. One such citizen jury in Mali in 2006 heard the testimony, deliberated, and then issued the predictable verdict: "Yes to traditional seeds! No to GMOs! We do not want GMOs in Mali at all" (IIED 2006). This ventriloquized verdict was taken seriously in the local and national media, and it played a role both in slowing progress toward passage of a national biosafety law in Mali and in blocking cabinet approval for field trials of *Bt* cotton.

European donors are also eager to finance the drafting of tight regulations against GMOs in Africa. In 2005 Germany's foreign assistance agency, GTZ, launched a three-year €2 million project to promote through the African Union the acceptance of a highly restrictive African "model law" that would effectively preclude the approval of any GM foods and crops in Africa (GTZ 2006). Earlier in 2003 the Norwegian government gave Zambia $400,000 to help that country enforce a GM-free policy. The bulk of this Norwegian money went to buy equipment and train scientists for a new Zambian lab dedicated to detecting GMOs both in the marketplace and at the border. The primary recipient of these funds was Dr. Mwananyanda Lewanika, the head of Zambia's National Institute for Scientific and Industrial Research and a leading proponent of the Zambian GMO ban earlier in 2002 (Ngandwe 2005). At the Norwegian end, the institution selected to train the Zambians was Terje Traavik's ultra-precautionary Institute of Gene Ecology. The goal of this assistance from Europe was not to help Zambians use GMOs, but instead to help them keep this new science out of the country.

More recently in 2006, DGIS from the Netherlands funded a similar effort in Zimbabwe by RAEIN-Africa to train twenty-six participants from eight countries in how to detect any presence of GMOs in the food supply (Tsiko 2006). The Dutch were apparently untroubled by the irony that their effort to protect Africans from the unproven risk of GMOs was being run from the local headquarters of Zimbabwe's Tobacco Research Board.

Programs by European donors to spread caution about GMOs in Africa routinely trump the various official efforts made to promote GMOs by the U.S. government. Since the 1990s, USAID has offered persistent support for GM crop research and commercial adoption in developing countries, including Africa, but with little effect. This U.S. effort initially centered around an Agricultural Biotechnology Support Project (ABSP) led by scientists at universities in the United States, but as regulatory hurdles grew, USAID began shifting its resources into policy assistance, via a new Program for Biosafety Systems (PBS) run through the CGIAR. The naïve hope was that African governments, after accepting USAID's money and training, would then go ahead with a full program of field trials and commercial approvals for GMOs. This did not happen. Only a few of the new GMO technologies funded by these USAID programs have advanced to the stage of field trials to test for biosafety, and none has yet been approved for commercial release (Spielman, Cohen, and Zambrano 2006). In 2004 the United States gave Nigeria $2.1 million to help develop an institutional infrastructure for approving GM crops, and in the following year USAID flew Nigeria's entire national biosafety committee to Missouri to show them how unthreatening GM crops actually looked in the field. Following this visit the Nigerian committee returned home, postponed several meetings, approved no field trials, and then asked for more money (Hand 2005).

The United Nations System

In addition to foreign assistance programs, a second formal channel for exporting European-style GMO regulations to Africa has been a global program on biosafety regulation run by UNEP and funded by the Global Environment Facility (GEF). This program coaches poor countries that lack their own GMO biosafety laws to embrace, in effect, a European-style precautionary approach. UNEP/GEF does this global work under the umbrella of the 2000 Cartagena Protocol on Biosafety, an agreement now ratified by 138 countries (including thirty-nine African countries) negotiated to govern the transboundary movement of *living* agricultural GMOs, called LMOs. The protocol primarily governs international trade in LMOs, but through its implementation-support programs UNEP shapes African policies behind the border as well.

The origins of the Cartagena Protocol trace back to Article 8(g) of the 1992 Convention on Biological Diversity (CBD) of the United Nations, which did require parties to take measures within their own borders to regulate, manage, and control risks associated with LMOs. Still, this original article allowed individual countries to decide for themselves how strict or lax their domestic regulations would be. This original flexibility brought objections from European environmentalists, who began pushing for an international negotiation to set in place a stronger global regulatory standard, something closer to the 1989 Basel Convention on the Control of Transboundary Movements of Hazardous Wastes. The likening of GMOs to hazardous wastes was a bizarre and inappropriate framing, but it was popular among environmental advocates in Europe and it was sold to Africans as something the UN was doing to protect their rich biodiversity.

It would be one thing if the protocol had been negotiated to cover any of the many real and documented risks to biosafety and biodiversity facing the developing world, such as bioinvasions from exotic wild species. Instead the protocol ignored these real problems and focused only on what was bothering European environmentalists at the time: agricultural GMOs. Still, the final language of the protocol, as agreed to in Montreal in 2000, was actually quite flexible, since the United States and a "Miami Group" of agricultural exporters had strongly opposed taking the same approach used in Basel. The final language of the protocol took care not to impose serious new trade restrictions on LMO farm commodities so long as they were destined to be consumed as food or animal feed, or in industrial processing. Only when LMOs were intended for release into the environment—for planting by farmers—did the protocol require burdensome steps such as advanced informed agreement (AIA) from the importing country. The process of negotiating and then implementing the Cartagena Protocol nonetheless gave European opponents of GMOs new opportunities to work against the technology in Africa.

Many African delegates originally came to these negotiations fearing not that GMOs were dangerous but that they might work so well in prosperous countries as to leave African agriculture further behind. The Africans were turned around quickly by scare stories about GM food safety and biosafety spread by groups such as Greenpeace, Friends of the Earth International, and the Third World Network, which had been granted access to the negotiation process. As explained by Bas Arts and Sandra Mack, "Generally, many developing countries had only a limited knowledge on biosafety issues because of lack of financial and scientific resources. It was the NGOs which made them aware of the (potentially) negative consequences of the transboundary movement of

GMOs for their countries, particularly for their rich biodiversity, traditional agriculture and indigenous people" (Arts and Mack 2007, p. 53). The formal record of the negotiation captures how poorly informed some African delegations in fact were, and how easily they could be frightened away from the new technology. It did not help that some had recently seen a 1993 Hollywood film called *Jurassic Park.* One African delegate to a working group meeting in 1996 warned his colleagues that modern biotechnology might bring back to life dangerous species that had been extinct for millions of years (ENB 1996).

Since the negotiation was framed as an international environmental agreement, most African nations gave their environment ministries responsibility for handling the job. These underfunded and poorly informed African environment ministries were easy for Europe's better-established environment ministries to influence. In this regard an important bond was formed between Britain's environment minister and chief negotiator, Michael Meacher, and the Ethiopian environment minister, Dr. Tewolde Gebre Egziabher, who ended up serving as chief negotiator for the African group of countries from 1996–2000. Because most African nations were completely unprepared for the negotiation, Tewolde assumed a leadership role, and once he had done so his African colleagues—in the spirit of continental solidarity—followed along.

Tewolde had been trained as a plant ecologist at the University of North Wales and held views on commercial agriculture that were of limited appeal to his fellow Ethiopians back home. He was an enthusiast of organic farming, an opponent of chemical fertilizers, and even a skeptic regarding agricultural irrigation (Tewolde and Edwards 2005). Together with Sue Edwards, an advocate for organic agriculture, and with support from an NGO based in Malaysia named the Third World Network, he had created his own NGO in Ethiopia in 1995 called the Institute for Sustainable Development (ISD), which also partnered with the Gaia Foundation in the United Kingdom. Because Tewolde's positions on agriculture resonated so strongly with critics of science-based farming in Europe, many Europeans promoted him as Africa's authentic voice on GMOs.

Throughout the Cartagena Protocol negotiations, Tewolde could be counted on to depict all of Africa as skeptical toward GMOs. In 2002 when controversies erupted over the delivery of GM food aid to Africa, Meacher brought Tewolde to London to offer a briefing at the House of Commons, where he told the MPs directly that GM crops were *not* the solution to Africa's hunger problems (Genetic Food Alert 2002). In 2003 Meacher brought Tewolde back to London to testify at a press conference at the Gaia Foundation that Africans did not want GM foods (Gaia Foundation 2003).

The protocol provided European activists opposed to GMOs with even greater opportunities for influence inside Africa in its implementation phase. Because this was a UN agreement, it was UNEP that took the lead in implementing the protocol, with funds it received in 2001 from GEF. With these funds, UNEP created a Global Project for Development of National Biosafety Frameworks (NBFs). Although the binding requirements of the protocol were actually quite weak, as noted earlier, UNEP was quick to assert that these formal requirements should be seen as a regulatory floor, not a ceiling. UNEP advised developing countries to take the European approach by setting in place separate standards for approving GMOs, and to hold back on approving GMOs until new biosafety committees were created to approve crops on a case-by-case basis according to these standards. UNEP also voiced its opinion that it would not be right to approve a crop simply because it had been approved somewhere else (even Europe), since each country had to have its own separate national regulatory capacity (UNEP 2006). Poor countries, UNEP asserted, should also operate their assessment and approval systems using Europe's precautionary standard.

The tool kit provided to governments in Africa by this UNEP project was a precautionary regulator's dream. It listed all of the activities governments might wish to regulate separately: the contained use of GMOs, field trials of GMOs, the commercial release of GMOs, the marketing of foods derived from GMOs, the transport of GMOs, and the packaging, labeling, and disposal of GMOs. The toolkit also encouraged poor countries to consider regulating GMOs for a range of goals that went beyond health and safety, such as sustainable development, the preservation of traditional agricultural practices, and the safeguarding of cultural or religious interests and values (UNEP 2005). UNEP promoted this expansive regulatory agenda by sending in experts—typically from Europe—to review draft regulatory legislation and hold workshops. It also sponsored visits to Switzerland, Germany, Netherlands, Belgium, and France to show officials in developing countries how a proper system should operate. Because European states contribute three times as much to the GEF Trust Fund as does the United States, it was natural to find this GEF-funded project pushing an essentially European line (GEF 2006).

UNEP's efforts to promote strict biosafety practices for GMOs were influential in Africa because governments there had a greater need for the "capacity building" money that GEF provided. Between 2000 and 2006, GEF spent $74 million to promote GMO biosafety in the developing world as a whole, much of it in Africa (UNEP 2006). UNEP was also particularly influential in Africa because most governments there had not previously addressed the issue of GMO biosafety, meaning they had no existing biosafety policies in place that might have to be challenged or altered. It is always easier to write on a blank slate. In countries with more advanced biotechnology capabilities, such as the Philippines, China, South Africa, and Argentina, UNEP's highly precautionary vision met effective resistance from local ministries of agriculture and ministries of science and technology, but in most of Africa, where national agricultural scientists were not yet working with GMOs, the precautionary vision was easy for UNEP to promote.

Of the twenty-three African governments that had completed an NBF by October 2006, all but one (the Republic of South Africa) began as "blank slate" countries with no previous legal regulations in place on LMOs. It is telling that twenty-one of these twenty-three countries completed the UNEP-GEF program by embracing the strongest possible approach, a so-called "Level One" approach in which LMOs would have to be regulated using binding legal instruments approved by the legislative branch of government (e.g., parliament). In the rest of the developing world, where governments had more independent knowledge about the science of GMOs and were less influenced by donors, this inflexible Level One approach had been wisely avoided. In Asia, of the eighteen countries that completed an NBF with UNEP by October 2006, only one followed the Level One approach. In Latin America, of the eight countries that had completed an NBF, only one followed the Level One approach (UNEP 2006).

UNEP claims its goal is to facilitate the safe use of GMOs, but its efforts have mostly led so far to an extended prohibition against any use at all. Of the twenty-three African countries to have completed an NBF, only one—the Republic of South Africa—has yet made it legal for any GM crops to be grown. South Africa escaped regulatory paralysis because it was more advanced and had a functioning biosafety policy of its own in place long before the UNEP program began operating.

The UNEP/GEF program provides a curious sort of assistance. It does nothing to enhance the productive capacity of African farmers; it only builds regulatory capacity. The only special skills enhanced by the program are risk-assessment skills and the only technologies transferred are for detection or testing, not technology development. In 2006, Uganda received from UNEP $50,000 worth of equipment to detect GMOs (a DNA concentrator, a thermocycler with accessories, a gel documentation/analysis system, a set of pipettes, and a microplate reader) to help regulators test for the presence of GM foods. Proud Ugandans bragged this would make them a "regional center of excellence" in GMO detection and assessment (Wamboga-Mugirya 2006). But the last thing needed in science-starved Africa is this kind of excellence at keeping productive new technologies completely under wraps.

UNEP now has a new project for discouraging the use of GMOs. Over the next four years it will promote the negotiation of a new international agreement to govern "liability and redress" in the event that the new Cartagena Protocol is somehow violated. The intended purpose will be to ensure that if an introduced LMO does "harm" to biosafety, then an individual, a firm, a country, or a multinational company will be found legally liable and forced to pay. Similar liability and redress measures were earlier demanded by activists opposed to GMOs in Europe, so it is unsurprising to see UNEP now exporting this added regulatory burden into Africa as well. Nor is it surprising that the leading African partner in this new policy transplantation exercise will again be the environment ministry of the government of Ethiopia, still led by Dr. Tewolde (IISD 2007).

While UNEP has been promoting the precautionary regulation of GMOs in developing countries, several other international organizations that might have been expected to promote the technology opted not to do so. Inside the World Bank, crop specialists at the technical level were initially interested in GMOs, and in May 1999 a biotechnology task force produced a draft discussion paper that noted considerable potential in the technology (World Bank 1999, p. 15). At a higher level, however, Bank leaders were not ready to begin lending for support of GMOs. This 1999 task-force report had come along just as political controversy over GM foods and crops was climaxing for the first time in Europe. After an interlude of indecision on what to do with GMOs, the World Bank finally found a way in August 2002 to postpone having to decide. The Bank launched a multiyear global consultation process designed to assess opinions about the new technology—along with opinions about all other agricultural technologies. It took more than two years just to establish the international governance structure for this sprawling $10 million writing project, which eventually was expanded to include twenty-nine co-sponsoring agencies, forty-five different governments, and eighty-six NGOs (World Bank 2005). In its 2008 *World Development Report* the World Bank noted that GM crops had shown considerable potential to help the poor, and conceded there was no scientific evidence yet of any new risks to human health or the environment. Instead of endorsing wider use of GMOs, however, the Bank stressed the need to cater to public anxieties by adopting strict, case-by-case approval systems that followed the "precautionary approach" (World Bank 2008).

The FAO, based in Rome, is another intergovernmental organization that might have been expected to endorse GM crops for Africa, since it had a long history of promoting technologically advanced commercial farming methods around the world, especially during the Green Revolution. Yet FAO's Senegalese director general, Jacques Diouf, did not at first embrace the technology, saying FAO's hunger reduction goals up to the year 2030 could be reached without GMOs. Eventually, when evidence of the safe and effective use of GM crops began to accumulate, FAO allowed a team of its own economists in 2004 to publish a report—which Diouf approved—summarizing this evidence and endorsing GM crops as a potential source of productivity and income gain even for low-resource farmers (FAO 2004). An outraged reaction came immediately from the international NGO community, which had never liked FAO's traditional protechnology bent. A coalition of 670 separate NGOs sent Diouf an open letter expressing their strong disagreement, calling the report a "stab in the back" to farmers and the rural poor (GRAIN 2004). Diouf defended the substance of the original report but as an act of contrition promised closer consultations with the NGO community in the future, and FAO as an organization has said little about the possible benefits of GMOs ever since.

The Consultative Group on International Agricultural Research (CGIAR) is another protechnology organization that has mostly tried to duck this controversy. European contributions to the CGIAR budget are now twice as large as contributions from North America, so the CGIAR system has been understandably cautious about promoting research on GMOs. As of 2007, only 7 percent of the budget of the CGIAR system (about $35 million) was being spent on any kind of biotechnology, and only 3 percent was going to work on transgenics (CGIAR 2007).

When CGIAR centers do occasionally become involved with GMOs it is usually as a low-profile partner in a project initiated and funded by others. In Kenya since 1999, the International Maize and Wheat Improvement Center (CIMMYT) has been operating with KARI on a project to develop GM insect-resistant maize for Africa. This project (IRMA) is funded by the Syngenta Foundation for Sustainable Agriculture and the Rockefeller Foundation. In the Philippines in 2002, the International Rice Research Institute (IRRI) began working with GM beta carotene–enriched Golden Rice, but only because funding was available from a number of outside donors. So sensitive is the issue of GMOs for the CGIAR that its annual report in 2005 made no mention of either of these projects, and when it launched an initiative in 2004 to develop biofortified crops with higher vitamin, iron, and zinc content, it said in advance no techniques involving genetic modification would be used for the first four years, and even after that—for the next six years—no GM crops would be released through the project (HarvestPlus 2004).

International Commodity Markets

African countries also get Europe's message to stay away from GMOs through the private international commodity markets that remain important to Africa for export earnings. Through these markets Europe tells Africa any move toward planting GMOs might put export sales at risk. Commodity exports are of diminished importance in much of the developing world today, but they are still highly salient for governments in Africa. According to one count, thirty out of forty-eight countries in Africa specialize in the export of nonfuel commodities, including agricultural commodities, compared to only two out of thirty-two Asian developing countries with this same dependence. Africa is also uniquely dependent on commodity sales to Europe. The European Union buys roughly €7 billion in farm products from Africa every year, six times as much as the United States buys (European Council 2005). The preeminence of Europe as an export destination for African food products reflects both geographic proximity and the substantial trade preferences the European Union gives to African goods as a holdover from colonial times. Europe's commodity imports from Africa are also more likely to be products headed directly to the food plate, such as meats or fresh fruits and vegetables, not just bulk shipments of coffee, tea, sugar, or cocoa.

When European consumers began expressing anxieties about GM foods in the late 1990s, it was only a matter of time before private importers in Europe would start sending signals to Africa that growers and exporters there should remain GMO-free. In 2000, private European buyers stopped importing beef from Namibia because the animals had been raised on yellow maize imported from South Africa, which could have been genetically modified. To regain their access to the European market, Namibian meat producers stopped buying GM feed maize from South Africa. The managing director of a Namibian company explained that in his business, "the consumer is, at the end of the day, the deciding factor" (Graig 2001). For African exporters, the most salient consumers are Europeans.

Fear of lost export sales also became an important factor in Zambia's 2002 decision to refuse imports of GM maize. Since 1998 a private international company in Lusaka named Agriflora Ltd. (it exports roses along with fresh vegetables) had been producing vegetables on a 200-hectare company farm for direct export to supermarkets in the United Kingdom, certified as organic by Britain's Soil Association (FAO 2001). In the summer of 2002, as Zambia was confronting its drought and the need for substantial international food aid, Agriflora received phone calls from British supermarkets explaining that exports of organic baby corn to the United Kingdom would be in jeopardy if food aid shipments containing GM maize were allowed into Zambia. In response Agriflora and other export-oriented growers asked President Levy Mwanawasa to reject the food aid. As Robert Munro, the general manager of vegetables at Agriflora, explained later, "Our main selling point is that Zambia is GM free" (Plotkin 2003). President Mwanawasa's official team of expert advisors later confirmed that exports were a concern, citing "a potential risk of GM maize affecting the export of baby corn and honey in particular and organic foods in general to the European Union if planted" (Government of Zambia 2002). The Zambia National Farmers Union (ZNFU), dominated by commercial growers of export products, also wanted a ban on imports of GMOs to protect foreign sales. In 2003 the president of the ZNFU told a meeting of international agricultural investors: "The food shortages of 2001/2002 season brought about the issue of GMO relief maize for Zambia. The Government made a stand not to allow GMO maize in the country. This has also been the stand of the ZNFU on GMOs. A major factor is the position of European importers that if Zambia adopted GMO crops, exports of crops such as tobacco, sweet corn, baby corn and organic products from Zambia will not be accepted" (Robinson 2003).

In reality the export risks Zambia and the rest of Africa might face from planting GM crops are small in magnitude, because GM crops such as soybeans and maize are not an important part of what Africa sells to Europe. The products sub-Saharan Africa does export to Europe, including coffee, tea, cocoa, sugar, mango, green beans, and banana, are not yet being grown anywhere in GM form, so it is unlikely Europe would shun them simply because an African country began planting GM maize. The only prominent African export to Europe commercially planted so far in a GM form has been tobacco, grown for a while in China in the 1990s. There is a comical possibility that some health-conscious smokers in Europe might refuse to buy GM tobacco from Africa, since health-conscious smokers in Japan did earlier refuse to buy GM tobacco from China. Yet the import or planting of GM maize in Africa is unlikely to compromise any significant value of export sales to Europe. In 1997 the Republic of South Africa began planting all the standard GM crops—cotton, maize, and soybeans—without any noticeable adverse effect on its lucrative export sales of fruits and vegetables to Europe. South Africa's export sales of fruits and vegetables both to Germany and to the United Kingdom more than doubled between 2000 and 2004.

Possible export sales losses have nonetheless been an understandable worry to governments in Africa. They all watched as corn growers in the United States began experiencing a significant annual $300 million loss in maize export sales to Europe in the wake of the *de facto* 1998 EU moratorium on new GM crop approvals. In May 2003 the United States, supported by Argentina and Canada, initiated legal proceedings in the WTO against this EU moratorium and eventually won a technical victory three years later when a dispute settlement panel found the EU moratorium violated WTO obligations, in part because it had not been based on scientific evidence of risk (WTO 2006). The European Union did partly restart its approval process for GMOs in 2004, but since six renegade member states (Austria, France, Germany, Greece, Italy, and Luxembourg) continued to impose a complete ban on seven crops fully approved by EU regulators before the moratorium, the door to the EU market was hardly reopened. U.S. trade officials had hoped their WTO victory would deter other importing states from imposing restrictions on GMOs without scientific evidence of risk, but African exporters took home a different lesson: don't start planting GM crops in the first place.

This has also been a lesson taken away by some American growers and exporters. In 2002 commercial exporters in the United States pressured private biotechnology companies not to put any GM rice varieties on the market for fear of lost export sales, a fear later vindicated in 2006 when traces from one of the withheld rice varieties were found in U.S. export shipments, triggering significant export losses. Upon learning the shipments were "contaminated" (to use the language employed by critics of GMOs), Japan put a block on all imports of long-grain rice from the United States, and the European Union began demanding preshipment testing. Approximately $150 million worth of sales were lost as private importers around the world began to shun all long-grain rice exports from the United States.

Genetically modified wheat has also been kept off the market in the

United States primarily to protect export sales. In 2004 Monsanto was finally persuaded by U.S. and Canadian growers and exporters to halt its plans to commercialize a GM wheat variety, because importers in Europe, Japan, and South Korea signaled they might not buy it. According to one estimate, if GM wheat varieties had been introduced into production, roughly one-quarter to one-half of the market for U.S. hard red spring and durum wheat exports would have been lost and prices would have fallen by one-third (WORC 2006).

Africans had other reasons as well to worry about the commercial implications of planting GM crops. Even if the varieties grown might be officially approved for import into the European Union, exporters of GMOs would have to comply with escalating EU regulatory demands, such as the new labeling and tracing rules set in place by the European Union in 2004, rules that require costly product segregation and record keeping all along the marketing chain. The American Soybean Association complained in 2006 that Europe's new labeling rules were costing U.S. farmers and food companies that handled GM products hundreds of millions of dollars a year in lost sales (ASA 2006). Africans can't afford such losses.

Campaigns by NGOs

International NGO campaigns against GMOs are a final source of external influence on Africa. For at least the past decade scholars have noticed the growing influence of "transnational advocacy networks" in pushing for policy change in developing countries (Keck and Sikkink 1998). Most of the changes these networks seek are highly desirable, but in the case of GMOs civil-society campaigners have been transplanting into Africa a European attitude toward the technology that fits poorly with local needs.

NGO campaigns have influence in Africa because the continent has come to depend heavily on NGOs, particularly since the decade of the 1980s when governments there were running short of money and found themselves obliged to cut back on public-sector service delivery, under debt burdens and the structural adjustment disciplines of the World Bank. Where the state bureaucracies withdrew, NGOs tried to step in. In Kenya by the end of the 1980s international NGOs had taken at least partial responsibility for two-thirds of all secondary schooling, more than one-third of all health care, and a wide variety of other activities in both urban and rural areas (Bratton 1989). Even the poorest countries in Africa now have direct ties to scores of international NGOs. The country in Africa with the *fewest* ties to such organizations, Chad, was linked as of 2000 with 190 different international NGOs (Beckfield 2003).

International NGOs like to work in poor countries through local partner organizations, providing them with money, vehicles, training, staff, and opportunities for international networking and travel. This relationship leaves the local partner dependent upon and deferential to the parent organization (Bradshaw and Schafer 2000; Postma 1994). International NGOs and their local partners nonetheless enjoy considerable credibility in developing countries, particularly in Africa, in part through their ability to claim, usually with some justification, that they are less corrupt, in some ways more democratic, and usually more concerned with the poor than local government bodies. Transnational NGO advocacy networks are known to be most effective when they can claim to be speaking on behalf of vulnerable groups that are unable to speak for themselves, such as imprisoned victims of human rights abuses (Keck and Sikkink 1998, p. 204). In Africa, NGOs become influential by claiming to speak for the rural poor, a claim difficult to challenge since the rural poor themselves are geographically isolated, illiterate, and generally absent during policy debates.

Ever since GM crops were first commercialized in the middle years of the 1990s, scores of international NGOs have joined in a broad transnational advocacy campaign against them, led by two formidable European environmental groups, Greenpeace and Friends of the Earth. Greenpeace International, based in Amsterdam, has local chapters in forty countries and employs over 1,000 full-time staff members. Friends of the Earth International, also based in Amsterdam, has chapters in sixty-eight countries and approximately 1,200 full-time staff. Consumer-protection NGOs are a secondary partner in these international campaigns against GMOs. Consumers International (CI), a global federation of more than 230 consumer-advocacy organizations concerned with food safety in 113 different countries, conducts an international campaign called "Consumers Say No to GMOs." Financing for these campaigns can come from many small, individual donations (the source of most Greenpeace money), but funding is also provided in bulk by governments, private foundations, or politically motivated individuals.

Today's campaign against GMOs in poor countries began as an extension of a campaign originally waged against the new technology in Europe, where NGOs played a powerful role from the start, advocating for tightened regulations through the European Union's Directorate General for Environment, Consumer Protection, and Nuclear Safety (known as DG XI) (Patterson 2000). In 1996, when GM soy and corn were first approved for planting and import into Europe, NGO activists responded with street demonstrations, efforts to block the unloading of ships carrying soybeans from the United States, and invasions of farm fields to uproot GM crops (Bernauer and Meins 2003). Once these protests produced the desired result in Europe, in the form of tighter labeling requirements and a 1998 moratorium on new GM crop approvals, the campaigners took aim at the rest of the world, including Africa.

One early international NGO victory against GMOs was the Cartagena Protocol, as noted earlier. The completion of the protocol in 2000 was important because it gave NGOs an international legal standard of precaution they could use as a reference point. For example, they could now begin attacking the United States for its longstanding practice of delivering GMOs in bulk shipments as food aid to poor countries. Under the protocol, the kernels of GM corn contained in these shipments were classified as LMOs, meaning the importing country was entitled to a warning label. The United States resisted providing such a warning, reasoning that the corn in these shipments was approved as safe and identical to what Americans had been willing to buy and consume for several years without any warning labels. Besides, the United States had never agreed to become a formal party to the protocol. In the eyes of the NGOs, of course, this implicit rejection of the protocol only made the United States a more inviting target for attack.

The self-serving manner in which the United States gave its food aid was also easy to criticize. Under the terms of a 1954 Act of Congress known as Public Law 480, United States food aid to poor countries had to be provided in the form of actual commodities purchased in the American marketplace. Economists can show it would be more cost-effective to purchase food for humanitarian relief either in foreign markets closer to the crisis or from foreign exporters closer to the final destination of the aid, but Congress will only continue to fund America's large food aid program (which by itself makes up more than half of all international food aid given) if assured that all the aid given will be purchased from American farmers and exporters. Since 1996 this has meant all American maize and soybean food aid has contained GMOs.

Led primarily by Friends of the Earth, the NGO community in 2001 began depicting unlabeled GM food aid from the United States as part of a stealthy scheme to dump surplus quantities of unhealthy American foods onto the vulnerable poor. Friends of the Earth first distributed test kits to its field offices to document the presence of the GM products in U.S. food aid shipments, finding GM corn and soybeans in some food aid shipments from the United States to Bolivia, Colombia, and Ecuador. This was then presented and publicized as evidence of "GMO contamination around the world," setting the stage for the African rejections of GM food aid that began in the following year (FoE 2001).

As I briefly mentioned in the Introduction, the 2002 rejections actually began in Zimbabwe in May, when the government in Harare decided to turn away a 10,000-ton shipment of U.S. corn for fear that the shipment was "contaminated" with GMOs. Officials worried that if any of the GM maize kernels were planted by farmers, national biosafety regulations would be violated and future export sales of hybrid maize could be put at risk. Zimbabwe eventually agreed to accept GM maize as food aid if the kernels were milled before delivery so they could not be planted, but in the meantime the rejected shipment of whole kernels was diverted to other needy countries in the region: Mozambique, Malawi, and Zambia.

Zambia had routinely accepted GM food aid shipments in the past, but criticism from the NGO community was making this more difficult by 2002. European NGOs that were critical of the technology had influential allies inside Zambia's policy elite. In 2002 Dr. Mwananyanda Lewanika, the executive director of Zambia's National Institute for Scientific and Industrial Research (NISIR), was developing a close relationship with a European NGO (Norway's Institute for Gene Ecology) that gave him reason to advocate against accepting food aid. Earlier in 2002, at a Cartagena Protocol meeting in the Netherlands, Lewanika had made preliminary plans to work with the Norwegians to operate a precautionary biosafety awareness program through his institute (Lewanika 2002). It was Lewanika who led in presenting the technical case against accepting GM food aid in Zambia. At a public meeting in August 2002, he alleged that GM foods could increase cancer risks and were contributing to a growing public health danger in the form of antibiotic resistance to infections. He then invoked the precautionary principle and called for his government to "take steps that will lead to the development of national capacity to detect GMOs in food and foodstuffs" (Phiri 2002).

The most impassioned public speaker against GM food aid at this important open meeting in Zambia in August 2002 had also been influenced by international NGOs. Emily Sikazwe, executive director of a local NGO called Women for Change (WFC), told her fellow Zambians at this meeting how important it was to say no to GM food aid: "Yes, we are starving, but we are saying no to the food the Americans are forcing on our throats" (Phiri 2002). Sikazwe's own local NGO had earlier been spun off from a Canadian NGO (Canadian University Services Overseas, or CUSO) that engaged in "biotech teach-ins" against GMOs back in Canada in partnership with Greenpeace (CUSO 2011). Sikazwe's local NGO received its funding from a number of Canadian church and peace organizations, the Swedish embassy in Lusaka, the Norwegian embassy, and the Danish foreign assistance agency DANIDA (WFC 2007).

Two religious NGOs in Zambia led by expatriate Jesuit priests from the United States also joined in the attack against GM food aid. The Jesuit Centre for Theological Reflection (JCTR) and the Kasisi Agricultural Training Centre, both located in Lusaka, had begun proselytizing against GMOs in 2000. Earlier in 2002 they had jointly commissioned a research paper titled, "What Is the Impact of GMOs on Sustainable Agriculture in Zambia?," recommending that Zambia's policy regarding GMOs "should be prudently cautious like countries in the European

Union" (JCTR 2002, p. 17). These two Jesuit organizations embraced a doctrine—never endorsed by the Vatican—that all living things, including plants, should enjoy a God-given right not to have their "genetic integrity" altered (Lesseps 2003). As a more worldly motive, trainees from the Kasisi Centre were employed by Agriflora, growing organic produce for export to Europe. Kasisi's slogan was: "Organic Farming: Keep Zambia GM Free" (REN 2004). Training fees at the Kasisi Centre were paid by yet another European organization, the Swedish Cooperative Centre (FAO 2001).

American officials tried to reassure the Zambians about GM maize by inviting a delegation of experts on a fact-finding visit to the United States. This approach backfired when the seven-person Zambian delegation also traveled to Europe to gather facts in the United Kingdom, Netherlands, and Norway, and met with Greenpeace, Friends of the Earth, the Soil Association, Genetic Food Alert, and the Institute of Gene Ecology. Greenpeace warned the visiting Zambians that their organic produce sales to Europe would collapse if the nation accepted the new technologies, and Genetic Food Alert warned of "unknown and unassessed implications" from eating GM foods. A U.K. NGO called Farming and Livestock Concern warned the Zambians that GM maize could form a retrovirus similar to HIV (Daily Telegraph 2002). Upon returning home, the spokesperson for this Zambian expert group, Dr. Lewanika, asserted that his own anxieties about GMOs had only been confirmed by the trip (Government of Zambia 2002).

Having helped turn Zambia against GMOs in the summer of 2002, the NGO campaigners (including by now a number from North America) shifted their attention to the September 2002 World Summit on Sustainable Development (WSSD) nearby in Johannesburg. In anticipation of this UN event a San Francisco environmental advocacy group, Earth Island Institute, had organized an unofficial "World Sustainability Hearing" running parallel to the summit. A number of internationally prominent critics of GMOs from wealthy countries were recruited to speak at this forum, including Frances Moore Lappe (who celebrated organic farming as Africa's best alternative), Percy Schmeiser (a Canadian farmer who warned about the dangers of GM seed patents), Dr. Mae Wan-Ho (from the United Kingdom, who warned that

GMOs in fast foods made their way into bacteria in human stomachs), Emiliano Ezcuera (from Greenpeace, who claimed that GM soy production was generating superweeds in Argentina), and Juan Lopez (from Friends of the Earth in Belgium, who charged that the regulatory frameworks for GMOs were insufficient, even in Europe) (EII 2002).

Friends of the Earth then joined with IATP and the World Development Movement from the United Kingdom to persuade 140 local African civil society representatives and organizations in Johannesburg to sign an open letter to the World Food Programme and the U.S. government protesting shipments of GM food aid. This letter circulated widely on the Internet as the voice of Africa on the issue of GMOs. It was filled with alarming yet totally undocumented charges:

> The safety of GM food is unproven. On the contrary, there is sufficient scientific evidence to suggest it is unsafe. GM food can potentially give rise to a range of health problems, including: food allergies; chronic toxic effects; infections from bacteria that have developed resistance to antibiotics, rendering these infections untreatable; and possible ailments including cancers, some of which are yet difficult or impossible to predict because of the present state of risk assessment and food safety tests. (Third World Network 2002)

The Johannesburg summit also gave international NGOs an opportunity to put anti-GMO words into the mouths of local African farm organizations. An organization called PELUM (Participatory Ecological Land Use Management), claiming to represent 160 civil society organizations in Africa, organized what it called a "Small Farmers' Convergence" on Johannesburg that included a four-day caravan by 120 farmers that set out from Lusaka. Funding for this pilgrimage came from HIVOS and NOVIB in the Netherlands, FOS-Belgium, MISEREOR in Germany, and Find Your Feet in the United Kingdom. At a press conference in Johannesburg at the end of their walk these farmers announced, "We say NO to genetically modified foods" (GFAR 2002). PELUM's chief African organizer for this effort (not a farmer himself) later told interviewers he didn't like GMOs because he had learned that, if eaten, they would change the genetic composition of the human body (IMM 2002).

These inflammatory and unsubstantiated charges against GMOs in Johannesburg eventually provoked a response from the USAID administrator Andrew Natsios, who lost his patience after being asked by one African minister from a Muslim country "if it was true" that GM maize contained pig genes (Natsios 2006b). Natsios spoke out, calling the NGO efforts "revolting and despicable," but this only raised the profile of the issue. Having baited Natsios into the arena, the NGOs were more than happy to amplify the dispute. A Greenpeace spokesperson replied that the United States was being "arrogant to tell the Zambians what food they must accept," and Peter Rosset of Food First said he thought the Africans "should tell the U.S. to go to hell" (Martin 2002; Gidley 2002).

Building on their efforts in Zambia and Johannesburg, the NGO campaign later took its message to a number of other African countries. In Kenya, a collection of NGOs had organized themselves into a Kenya GMO Concern Group (KEGCO), and in 2004 two of its foreign-funded members, PELUM and ActionAid, led this coalition in a media campaign against passage of a draft biosafety bill, legislation that might lead to the commercial planting of GMOs in Kenya (PELUM 2004). For this campaign PELUM produced an article titled "Twelve Reasons for Africa to Reject GM Crops," a document that said GM crops were a threat to human health—and then invoked Europe's precautionary principle. The local Kenyan author of this document did not hide his deference to European opinion on the issue of GMOs: "Europe has more knowledge, education. So why are they refusing [GM foods]? That is the question everybody is asking" (Hand 2006).

Also in 2004 Friends of the Earth launched an African regional campaign to "challenge the myth of GM crops as a solution to hunger and poverty," working specifically to dissuade Angola and Sudan from accepting GM food aid (FoE Africa 2007). The Angolan government went along with this advice, rejecting GM maize in an unmilled form as food aid just when WFP was being forced to cut its normal feeding rations in the country due to funding shortfalls (Scott 2004). In Sudan, Friends of the Earth then led a group of sixty NGOs accusing the United States and WFP of coercing that country into accepting GM food aid (ACB 2004). Friends of the Earth also worked against GMOs in West Africa, holding a 2005 conference in Nigeria that brought together GM food critics from nine different African countries to demand "a complete moratorium on GMOs in Africa until their safety for our environment, health, and socio-economic conditions is established beyond doubt" (FoE 2005). Another Friends of the Earth conference held in Nigeria in 2006 called for an "immediate recall from Africa of all long-grain rice imported from the United States unless proven not to be contaminated" (FoE 2006). At this last meeting one spokesperson for Friends of the Earth said, "We refuse to be used as guinea pigs in big business's experimentations" (Investor's Business Daily 2006).

The South Africa Exception

If GM crops and foods have been kept out of Africa largely due to external influences from Europe and from other wealthy countries, how have they been able to go forward in the one case of South Africa? The Republic of South Africa has been as much a target for these external influences as the rest of the continent, yet it has officially approved both the consumption and planting of multiple GM foods and crops. How did South Africa manage to maintain its policy independence?

This was largely an accident of timing. South Africa made its most important policy decisions on how to regulate GM crops and foods early in the 1990s, before Europe had become alarmed about the technology. South Africa was far ahead of the rest of the African continent in part because its location in the southern temperate zone had made it an attractive place for international biotechnology companies to conduct off-season field trials of early GM crops. As early as 1979 the white Apartheid government in South Africa had begun readying itself for such trials by creating a South African Committee for Genetic Experimentation (SAGENE), which promulgated regulations and guidelines. Field trials of GM cotton plants began in South Africa in 1989, and commercial production of both GM cotton and maize was approved in 1997. South Africa therefore had its own regulatory policies and its own tested and approved GM crops already in the field a year before the 1998 EU moratorium on approvals for GMOs, three years before the 2000 negotiation of the Cartagena Protocol, four years prior to the

2001 creation of the UNEP/GEF special program, and five years prior to the inflammatory NGO campaigns conducted at the 2002 WSSD in Johannesburg. The rest of the African continent was more vulnerable to these influences due to its much later start working with the technology.

Once South Africa began commercializing GM crops in 1997, it never looked back. It passed a new GMO Act in 1999 and then commercialized additional varieties of GM cotton in 2000, soybeans in 2001, and white maize—a staple food crop—in 2002. During this later period international activists campaigned hard against the government's pro-GMO policy, working through a local NGO named Biowatch, which is largely funded by the German assistance agency GTZ (Wolson 2006). This campaign had only limited impact, however, because South Africa's large commercial farmers—still mostly whites—were by then already planting GM seeds profitably and with no apparent ill effects either to human health or the environment. In addition, since private industry (both national and multinational companies) stood to gain significant profits from commercial seed sales in South Africa, they fought back against Biowatch with an even better-funded propaganda campaign of their own, forming an NGO named Africabio that mobilized activists in favor of GMOs both from academia and industry. A comparably strong industry defense of the technology has not been mounted in the rest of Africa because there is less potential for lucrative seed sales, fewer local researchers to speak up for the technology, and because farmers have not yet been able to learn from experience that the technology works and is safe. In the poor countries of Africa GM crops are still confined to heavily regulated laboratories and tightly monitored field experiments, making them seem threatening rather than promising.

Why didn't South Africa's successful deployment of GM foods and crops create a spillover to the rest of sub-Saharan Africa? Here we find that cultural, climate, and technical factors all made a difference. South Africa is a temperate zone country, and most commercial farm operations are large and still in the hands of whites. Governments in the rest of Africa, whose rural citizens are smallholder farmers growing tropical varieties of staple food crops, do not usually look at South Africa as their farming model. In addition, the one GM food staple crop grown in South Africa, GM white maize, is not easy to spread illicitly from farm to farm in the rest of Africa because it is designed for the temperate zone, not the tropics, and because it is a hybrid variety that only performs well when seeds are bought anew every year from a commercial dealer with access to the parent lines—which means a dealer inside South Africa.

How Strong Are the Constraints?

The assertion here that external influences, mostly from Europe, have been keeping GMOs out of Africa leaves open the question of how strong those influences really are. What if a new GM crop with precisely the trait needed by Africa's poor farmers were to emerge from the research pipeline? For example, what if agricultural scientists developed GM crops better able to tolerate drought? Perhaps this would be a crop trait so compelling in Africa as to shift the political dynamic and unblock the technology. Perhaps the European donors and NGOs that have so far opposed GM crops because they find the first generation of traits noncompelling would think again about genetic engineering if it could deliver crops to Africa with improved drought resistance. In the next chapter I discuss how genetic engineering is now in the process of developing and delivering crops with improved drought-stress tolerance, making possible a preliminary test of this important question.

5

Drought-Tolerant Crops—Only for the Rich?

In Africa, rainfall is destiny. Only 4 percent of cropped area in Africa is currently irrigated, so nearly all farmers depend solely on what falls from the sky. Roughly 40 percent of farmers grow crops in arid or semi-arid regions marked by long dry seasons and scant rainfall even in the wet season. For these farmers, dryness in the field can set a strict limit on the productivity of their labor. Dryness determines which crops they can plant and how well the crops perform, and cyclical drought is a constant threat to income. If a farmer decides, encouraged by a few wet years, to purchase seeds and fertilizer to expand crop production, the investment can be wiped out if the rains then fail, come too late, or don't last long enough. When crops fail due to drought, income falls, and expenses must be covered either by borrowing, if credit can be found, or by selling off household assets. Thanks to international food aid, drought-induced crop failures in Africa are less likely than in the past to induce famine, but dryness in the fields still brings to Africa's poor a damaging decline in income and wealth. Dryness in the fields deepens and perpetuates poverty.

Although all agricultural crop plants need water to grow and produce fruit, some need less than others. On most of Africa's savanna lands it is too dry to plant a thirsty crop like maize, but farmers can grow more drought-resistant crops such as sorghum or millet. The disadvantage of these crops is that they offer lower yields (compared to maize) and take longer to mature, which prevents double cropping (McCann 1999). It would be ideal if some of the drought-tolerance traits found in low-value crops such as sorghum or millet could be incorporated into higher-value field crops such as maize—or cotton, or groundnuts, or upland rice. If crop scientists could solve this problem, the productivity of farm labor in much of Africa would increase and farm income losses from cyclical drought would diminish. For such reasons plant breeders have at times referred to drought tolerance as the "holy grail" of crop science.

It was originally believed that genetic engineering techniques would be poorly suited to improving the capacity of plants to withstand abiotic stress factors such as drought, salinity, or soil toxicity, as these were assumed to be traits controlled by multiple genes. The insertion of a single *Bt* gene into a plant might confer resistance to caterpillars, but increasing drought-stress tolerance would probably be a multi-gene problem requiring conventional breeding techniques that select for the performance of the whole plant (Biotechnology and Development Monitor 1994). These early assumptions were undone in 1998 when a scientist at the University of Toronto named Peter McCourt managed to isolate a single gene that could help control drought tolerance in plants. He found that when the *ERA1* gene is suppressed, the stomatal aperture on leaves shuts down, helping to reduce water loss (Pei et al. 1998). A Canadian company named Performance Plants, Inc. saw commercial potential and began collaborating directly with McCourt, and other companies as well then began to probe genetic engineering approaches to drought-stress tolerance.

Additional discoveries were soon made. In 2003 the Monsanto Company announced that its efforts to find individual genes conferring "environmental stress tolerance" had been successful, and that it was testing several new transgenic varieties of drought-tolerant (DT) corn and soybeans (Monsanto Company 2003). In 2004 the Pioneer-Dupont Company presented to the media arresting photographs comparing two of its hybrid yellow corn varieties grown under identical drought stress, one containing an added gene that resulted in a visibly higher final yield (Oestreich 2004). Syngenta Biotechnology announced that it too was part of the race to develop DT varieties of corn, and would be employing the rDNA technology pioneered earlier by Performance Plants.

Prosperous commercial growers of hybrid yellow maize in temperate zone countries stand to make significant gains from this new advance in genetic engineering. Within only a few years they are likely to have at their disposal (if they live in a country willing to approve GM seeds) a new technology capable of reducing one of their more significant risk factors. Yet it is poor maize farmers in eastern and southern Africa who need this technology the most, since their exposure to drought risks is more threatening, and there is currently no guarantee they will get it. The companies now developing DT crops have little incentive to invest in moving their newly discovered DT genes into the tropical crops grown by poor smallholder farmers in Africa, who lack purchasing power and are not good seed customers.

The task of delivering DT crops to the poor should be one for public-sector foreign assistance agencies like the U.S. Agency for International Development, working with international financial institutions like the World Bank and the international agricultural research system of the CGIAR. Yet when given an initial opportunity to invest in this project in 2004–05, these institutions hesitated. To some extent this was because science-based farm productivity growth was no longer a priority for them, but equally important was the fact that the new DT crops would be GMOs, and thus likely to encounter political resistance and regulatory blockage.

In this chapter I review the unique constraint low rainfall and drought have placed on Africa's rural poor and the progress recently made in genetic engineering toward developing drought-tolerant crops. Then I describe the initial reluctance shown by donors in wealthy countries in 2004–05, when asked to finance a project that might have developed and delivered newly available DT crop technologies to Africa's poor. I examine, finally, the prospect that a new philanthropic organization—the Bill and Melinda Gates Foundation—will eventually be able to rectify this bad start.

Africa's Drought Challenge

In much of Africa there is no rain at all for months at a time during the "high sun" season, but then a shift in wind direction brings in seasonal moisture, and in only two weeks the landscape turns from parched brown to brilliant emerald green. Farm field preparation and seed sowing must be timed precisely to the arrival of these seasonal rains. If the rains fail to arrive on time, stop too soon, or fall too violently in a flood, the planted crops will fail.

Over the continent of Africa moisture patterns are highly diverse, with average annual rainfall as low as 5 mm in the northeastern Sahara (in upper Egypt it does not rain at all) and as high as 2,000 mm in the equatorial center of the continent, or 2,400 mm in some highland sections of the Great Lakes region just south of the Equator in the East, or even 4,000 mm along the extra-humid West African coast just north of the Equator. The equatorial center of the continent has no dry season, north and south of that center there are two dry seasons and two rainy seasons, in the hot Sahel region just south of the Sahara and in the southernmost regions of the continent there is only one rainy season, and on the Sahara itself, plus the Namibian desert and on the Horn of Africa, there is only a dry season (Nicholson 2000). Where topography is irregular in Africa, distinct microclimates abound. In much of arid Kenya no crops can grow for lack of rainfall, but moving up the slope of Mount Kenya in the highlands we first find just enough precipitation for planting sorghum, then for maize, then tea and coffee, then Alpine dairy farms, then a misty rainforest, and eventually a dramatic, snow-capped rocky summit 5,199 m above sea level.

Rainfall patterns in Africa are also irregular year by year and over longer periods up to decades or even millennia. Rains can be inadequate for two, four, or even six years in a row. Particularly in West Africa, rainfall fluctuations often last seven years or longer. The famine-inducing Great West African Drought of 1972–74 had its origins in a longer pattern of below-average rainfall that began in 1967 (Derrick 1977). Rainfall fluctuations that are centuries long have been noted as well: between 1600 and 1860 in West Africa average rainfall declined enough to dry up the important grazing lands south of the Sahara and force livestock herding communities to relocate 200–300 km to the south.

Dramatic as they are, such highly pronounced patterns of rainfall fluctuation are not the source of Africa's unique drought risk, since many of the world's regions are actually more prone to drought than Africa. Although roughly 20 percent of Africa's total agroecosystem is at high risk for experiencing drought, this is less than in either Latin America (31 percent), China (27 percent), or India (24 percent), and it is substantially less than for the United States (44 percent) or Eastern Europe (56 percent). Africa's drought challenge comes not from the greater variability of its rainfall but from its lack of any technological protection against those variations. Only 4 percent of cultivated land in sub-Saharan Africa is currently protected with irrigation (either diverted surface water or lifted groundwater) compared to roughly 40 percent of cropland now in South and East Asia. On these unprotected drylands in Africa it does not take much of a dip in seasonal rainfall to damage a crop. Perhaps the best measure of Africa's unique exposure to drought risk is the percentage of its agricultural production taking place in regions that are simultaneously warm rather than cool or temperate, arid or semi-arid rather than humid or sub-humid, and nonirrigated. Overall, 41 percent of Africa's agricultural production comes from lands that fall into this hot, dry, and unprotected classification, compared to only 26 percent for India, 13 percent for Latin America and the Caribbean, just 3.2 percent for the United States, 0.4 percent for Eastern Europe, and actually 0.0 percent for China (Pardey et al. 2006).

The irregular topography of much of Africa makes irrigation difficult. Bad roads and poor power infrastructures only add to the historically high engineering and construction costs. The large irrigation schemes constructed in the 1970s in the Volta Basin in West Africa cost an unaffordable $45,000 per hectare, some part of that undoubtedly swallowed by swindling and corruption (van de Giesen et al. 2005). Also, given the relatively thin population density of many of Africa's farmlands, irrigation investments can be hard for governments to justify on a per-capita basis. These investments are also resisted, not always with justification, on environmental grounds. Small-scale irrigation options such as rainwater harvesting, bucket irrigation, treadle and pedal pumps, and construction of small earthen dams may seem more affordable, but these systems can carry excessive labor costs in rural Africa.

Vulnerability to Drought as a Source of Poverty

We often think of drought in Africa as a cause of famine, but currently it is better understood as a cause of poverty. Africa's unique vulnerability to drought first attracted global attention in 1972–74, when several years of low rainfall in the Sahel region triggered a famine that led to an estimated 300,000 human deaths. Many who died were herdsmen who had waited too long in the 1972–73 dry season to move their cattle southward after the rains failed and the pasture dried out. Media images of the dying animals and extreme human hardship shocked the world. Then just a dozen years later in the mid-1980s a much wider drought spread through eastern as well as western Africa north of the Equator, forcing over 10 million farmers to abandon their lands. This time, perhaps a million people died. Outsiders again watched on television, and tried to improvise a humanitarian response. The emergency was worsened in many countries by armed conflicts that complicated the task of delivering food aid, yet an unprecedented $1 billion relief effort did save millions from starvation (FAO 2006).

Having learned from these devastating emergencies, the international community resolved to put in place for Africa a "famine early warning system," based on regular assessments of local rainfall patterns and market prices, to ensure more timely responses to future drought crises. This system performed exceptionally well in 1991–92 when a devastating rainfall failure in southern Africa cut aggregate cereal production in that region by more than 50 percent on average. In Malawi, Namibia, Swaziland, and Zimbabwe cereal production actually fell 60–70 percent. Local maize stocks had earlier been depleted, so an alarming 17–20 million people were at starvation risk, yet thanks to an effective food aid response, actual famine deaths were reported only in Mozambique, where aid had been impossible to deliver because of an ongoing civil war. In the rest of the region per-capita food aid deliveries were increased quickly and dramatically, from an average of less than 10 kg per person in the 1980s to more than 25 kg per person by 1992, a delivery of food that saved millions of lives (Pinstrup-Andersen, Pandya-Lorch, and Babu 1997). The timely nature of this response was facilitated by an improved Global Information and Early Warning System (GIEWS) run by FAO, a strong Famine Early Warning System Network (FEWS-Net) funded by USAID, and close coordination all the while between international food aid donors, the UN's World Food Programme, and assisting NGOs, plus a committed performance by the governments in the region.

When drought returned to southern Africa in 2001–02, this time pushing 15 million people into an increased dependence on international food aid, once again the response was so effective that essentially no famine deaths occurred. Despite the humanitarian alarm caused by Zambia's refusal to accept GM food aid during this drought, replacement supplies of non-GM white maize were successfully found in South Africa, so again there was no famine, not even in Zambia. Some journalists who had flocked to Zambia to collect graphic images of human starvation were even said to be disappointed when they found the disaster had been averted. One jaded correspondent was heard to complain over his dinner in Lusaka, "Where are the skellies? I don't see any, do you?" (Plotkin 2003).

These have been humanitarian assistance triumphs, but they tend to obscure the impoverishing impacts of drought that even emergency food aid cannot fully address. Starvation is avoided, but impoverishment is not. When food crops fail in a drought, the rural poor who can no longer provision their families from what they produce are obliged to purchase food in the marketplace with cash at a time when food prices (even with food aid deliveries) will be higher than usual. The poor will not have funds on hand to make these food purchases because their cash crops (for example, cotton) will have failed in the drought. At this point they will have to fall back on various short-term coping strategies: borrowing money (at a high rate of interest since many others are trying to borrow), selling off household assets (at a low price because others are also selling), cutting back on nonfood expenses such as school fees for the children (disinvesting in the future of their family), shifting labor into risky nonfarm activities (such as crafts, or panning for gold), or eating fewer meals a day (reducing their own capacity for productive work). Foraging for wild foods in the bush is sometimes also an option. Employing these diverse strategies rural households can survive several successive years of below-average rainfall, especially if they have access to some food aid (Adams, Cekan, and Sauerborn 1998). Yet they will emerge from the experience with fewer household assets, with more debt, in poorer physical condition, and with children who have fallen behind in school.

When the 1991–92 drought hit Namibia and cereal production declined by 71 percent, no human deaths were attributed directly to the drought, but many household assets were wiped out. Livestock owners were among the most devastated; 2,000 large domestic animals and 11,000 small farm animals died in one month alone, December 1992. Farmers whose maize and millet harvests failed were also badly hurt, as they were now obliged to purchase their food from shops. Some cut their food costs by sending the children away to be fed by grandparents who were still receiving a pension every other month, or by replacing their usual diets with wild berries and water lilies. Yet to raise the cash still needed they began selling off livestock, bicycles, radios, cooking pots, and in a few cases even their productive assets: farming tools and ploughs. The wise but sad rule among Africa's poor when selling off assets in a drought is, "goats before plows" (Devereux 1993; Devereux and Naeraa 1996). Droughts of this kind hit Namibia every three years or so.

Similar stories are common for all the drylands of Africa. In the South Wollo zone of northeastern Ethiopia, 80 percent of rural farming households earn less than $50 per year, and periodic drought is one reason they remain trapped in deep poverty. Oxen are essential for plowing the difficult soils in this region, and it usually takes a hardworking farmer ten to twelve years, assuming good rains, to earn enough to acquire an inventory of four or five oxen and sheep, which is considered the minimum needed to move above the poverty line. A drought can interrupt this asset-building process at any time. In a 1999–2000 drought in South Wollo, small farmers making slow progress in acquiring assets suffered a setback that increased the share of households in poverty (with fewer than four or five animals) from 60 percent to 78 percent. The rains eventually returned, allowing the poverty rate to fall back down to where it had been earlier, but only for as long as another drought could be avoided (Little et al. 2004).

Population resettlement away from dry areas is a strategy that has worked in some countries, but it is seldom an option in Africa. In the United States between 1931 and 1939 an eight-year drought hit the high plains of western Kansas, Oklahoma, and the Texas panhandle. This region, once properly known as the Great American Desert, had imprudently been put to the plow for wheat production during a period of higher rainfall. But then the rains began to fail, and the exposed topsoil blew away, creating a disaster area the size of Pennsylvania, known as the "Dust Bowl." Many of the farmers who lost their crops and livestock to this drought (or *drouth*, as they say in Texas) had already accumulated unpayable debts, so they tried to cope by selling off their animals and possessions. When these assets were gone they turned to collecting glass bottles for bootleggers (10 cents for a bucket load), selling skunk hides ($2.50 apiece), swapping a half a day's labor for a meal, or scavenging through a neighbor's garbage (Egan 2006). Eventually they swallowed their pride and stood in line to get public relief. But the dry conditions persisted, so in the end many decided to abandon their homes and head either back East, where the rains were more reliable, or out West to California, where migrant work could be found picking tomatoes, cotton, lemons, or peas. One out of three residents living at the center of the Dust Bowl eventually migrated, and many of the plowed fields returned to grassland pasture.

This relocation strategy was available to the Dust Bowl farmers because they were small in number compared to the nation's vast labor markets. In Africa the number of farmers vulnerable to drought makes this kind of relocation impossible. Recall that 40 percent of all production in Africa comes from hot, dry, and nonirrigated lands prone to drought devastation. Because of high rural population growth and low cropland productivity, farmers in Africa are not abandoning drought-prone lands; instead they are still moving in. Drought does induce significant migrations in rural Africa, but these tend to be local rather than distant, short-term rather than long-term, and circular rather than progressive or permanent (Findley 1994).

Because of these circumstances in Africa, crops better able to tolerate drought in a bad year are more valuable than crops able to give high yields in a good year. High-yielding crops are typically difficult for the poor to use because they perform well only with significant fertilizer applications, which may be unaffordable or even unavailable in the poorest rural communities. Moreover, on nonirrigated lands high-yield crop varieties will turn instantly into low-yield crops if the rains fail. On nonirrigated lands, high-yield crops result in yield instability, which can be a problem even when the rains are exceptionally good. Unusually bountiful harvests in good years can overwhelm local storage and transport capacities, flooding rural markets with so much produce that prices drop below what farmers need to earn to recover their seed and fertilizer investments. Where rural road and marketing infrastructures are poorly developed, producing too much in a wet year can be almost as bad for the poor as producing too little in a dry year.

This was one lesson learned from a Sasakawa Global 2000 program to promote high-yielding hybrid maize production in Ethiopia in the 1990s. When local farmers on several thousand separate plots of land were given access to hybrid maize seeds and chemical fertilizers, average crop yields quickly increased to a level more than 70 percent above the local average with traditional seeds and no chemical fertilizers (Howard et al. 1999). As this project scaled up, however, marketing bottlenecks led to a collapse in the local price for maize, a setback that eventually drove Ethiopian farmers away from the project and discouraged any subsequent use of hybrid seeds and fertilizers. The chairman of one participating farmer cooperative said, "We have increased our yields, but what do we do with it?" (Kilman and Thurow 2002). Crops improved with drought-tolerance traits will not lead to such problems even assuming the worst rural road and marketing infrastructures, because DT crops stabilize farm yields, and more yield stability translates into less risk and more income stability, and eventually greater asset accumulation.

Drought-tolerant crops will be of increasing value for Africa in the decades ahead, assuming a continuation of human-induced climate change. Climate scientists from the Intergovernmental Panel on Climate Change (IPCC) foresee that a continued warming will, as early as 2020, expose between 75 million and 250 million Africans to increased water stress (IPC 2007). A continued warming of the Indian Ocean will make the seasonal winds that bring rain to eastern and southern Africa 10–20 percent drier by 2050, compared to averages for the second half of the twentieth century (Kigotho 2005). The rate of soil moisture evaporation will in the meantime increase as the air warms (World Bank 2000). In November 2006 the United Nations projected that 600,000 square km of agricultural land in sub-Saharan Africa currently classified as moderately water constrained would likely become severely water limited in the future, assuming continued climate change (IFPRI 2006). William Cline has projected that because of climate change the total agricultural capacity of Africa (excluding Egypt) will decline by roughly 19 percent between now and 2080 (Cline 2007). These projected climate change effects make the development of crops better able to tolerate drought an even more obvious imperative in Africa.

Developing Crops Better Able to Tolerate Drought

Even in prosperous countries, crops better able to tolerate drought would be an attractive option with considerable commercial value. In some hot and dry parts of the United States (including parts of the former Dust Bowl), farmers are currently growing corn under costly irrigation and would welcome plant varieties that might require three or four fewer irrigation applications during a growing season. Farmers in more humid regions who grow corn without irrigation would pay more for varieties able to do better during interludes of heat and drought. For such reasons private seed companies such as Pioneer have for many years been using conventional breeding techniques to develop improved tolerance of drought stress in corn, selecting from research plots the individual corn plants that tend to do best when irrigation water is withheld. The challenge has always been to improve the performance of plants under drought conditions without diminishing performance under optimal conditions.

The pursuit of DT crops through conventional breeding can run up against several natural barriers. There may be too limited a range of genetic variation available within a given plant species and its sexually compatible relatives. Time delays will be encountered waiting for plants to mature, before the next breeding selection can be made. Moreover, crude plant selections based on any one trait can result in the loss of other traits that are no less important to the agronomic value of the crop. Today, advances in genomics and genetic engineering are breaking down some of these constraints. Modern genomics studies the traits of plants at the level of their fundamental genetic determinants (so-called quantitative trait loci, or QTLs), and QTL mapping studies are now isolating a number of phenotypic traits linked to drought-stress tolerance, such as osmotic adjustment, relocation of stem reserves, leaf senescence, and root architecture. In addition, marker-assisted selection (MAS) can speed the process of conventional breeding by allowing the results of a cross to be known without having to wait for a plant to mature. Genetic engineering now adds still more potential, because it allows desired genetic traits to be brought in from plants not sexually compatible with the crop being improved (Tuberosa and Salvi 2006). Genetic engineering can emerge as a technique of choice if there is limited phenotypic variation within a species and if the desired traits are under the control of a single gene, or only a small number of genes, available for transfer (Dale and Henry 2003). Fortunately, the improvements genetic engineering can provide will usually be additional to those delivered by other methods, so most often the optimal scientific approach will be to combine genetic engineering with modern breeding by marker-assisted selection.

As already noted, the first breakthrough in using genetic engineering to enhance drought tolerance came in 1998 when Peter McCourt of the University of Toronto isolated a gene in *Arabidopsis* that, when suppressed, made a plant's leaf pores super-responsive to dryness, thus controlling water loss. This discovery was made by accident when a careless graduate student forgot to water some genetically engineered *Arabidopsis* plants in a laboratory as he left for a long weekend, but then found them still alive and doing well upon return. McCourt learned he could produce this same effect in canola (a plant closely related to *Arabidopsis*) by placing a second copy of a transgene into the plant inserted backward as a "negative gene" that would be turned on only when the plant was short of water (Pratt 2005). Three years of extensive field trials demonstrated that these genetically engineered canola plants consistently out-yielded control plants by up to 26 percent under various water stress conditions, and with no yield loss under optimal conditions (Performance Plants 2005). McCourt's corporate collaborator, Performance Plants, began working to develop this same Yield Protection Technology (YPT) in corn, soybean, and cotton.

A different transgenic approach has been taken by researchers at the University of Connecticut, who in 2005 published a paper showing they had been able to engineer tomato plants for greater drought resistance by transferring in an *AVP1* gene from *Arabidopsis*, enabling the tomato plants to produce more of a specific enzyme called H+-pyrophosphatase, which made the roots better able to take up water during drought interludes. When subjected to a thirteen-day period of water deficit, the transformed tomato plants not only recovered more effectively when water was restored, compared to a control group; they also showed greater vegetative growth, flower set, and a higher yield under normal conditions (Park et al. 2005). This Connecticut team has subsequently attempted a similar result in rice, poplar trees, and legumes (Herwig 2006).

Outside the United States, the Republic of South Africa's Agricultural Research Council (ARC) as early as 1999 had found through genetic engineering a means to improve the drought tolerance of soybean plants, using a *P5CR* gene from *Arabidopsis*. After several years of laboratory and greenhouse tests, six genotypes were planted in a "rainout" shelter, and four of the six lines produced higher yields under drought conditions. In 2004 Egyptian researchers at Cairo's AGERI published results showing they had made wheat plants more tolerant to drought by inserting a *HVAI1* gene from barley (Sawahel 2004). In 2007 a Chinese research team collaborating with Purdue University successfully transformed cowpea plants with an *ABO* drought-tolerance gene, and also with an SOS gene for salt tolerance.

The number of genetic engineers currently working on DT traits continues to grow, but given the high regulatory costs of bringing any entirely new GM crop to the market, the first commercialized version of an engineered DT crop will almost certainly come from one of the three big biotechnology companies now in pursuit of this objective in the United States: Syngenta, DuPont/Pioneer, and Monsanto.

Syngenta Biotechnology in North Carolina is working in close cooperation with Performance Plants, Inc., the original Canadian developer of Peter McCourt's YPT, to enhance the drought tolerance of corn plants. Syngenta is combining McCourt's gene-insertion method with molecular-assisted breeding based on techniques such as high-throughput genotyping, precision phenotyping, geographic information systems science, and various tools from computation biology (Johnson 2006).

Even further along is Pioneer Hi-Bred International, headquartered in Iowa, a major company with $2.7 billion worth of annual crop-seed sales and traditionally a leader in corn genetics. Pioneer is now developing DT corn plants using marker-assisted breeding and map-based cloning plus genetic engineering. In decades past the company had employed a crude "drought house" with a roof but no sides near its Iowa research campus to help in finding corn varieties able to tolerate dryness, but now it operates a sophisticated drought research station in Woodland, California (where average rainfall during the growing season is less than one inch). At this research station, scientists can control water stress on plants over a wider area through precise irrigation. Pioneer also has a parallel facility in an arid region south of Santiago, Chile, so the company can conduct large-scale drought research on corn throughout the year (Wenzel 2006).

Companies tend to be secretive about any commercial research still in progress, but since 2004 Pioneer has posted photographs on its website showing its ability, with the addition of a single gene, to give its hybrid varieties of yellow corn significant protection from drought damage. Pioneer has recently been following two different approaches to drought-stress tolerance in corn, and both are still in an early "proof of concept" phase. As of 2007 the company was forecasting that its transgenic DT corn hybrids would not be on the market until sometime between 2012 and 2014. Pioneer has reassured its seed customers that the increased drought protection this technology offers will not compromise its maximum yield potential under wet conditions (Pioneer 2007).

In contrast to Pioneer with its strong legacy of conventional plant breeding, Monsanto moved into crop science more recently from chemicals, so it has favored using modern genetic engineering methods from the start. It was Monsanto that launched the GM crop revolution in the United States by commercializing herbicide-tolerant "Roundup Ready" GM soybeans in 1996. This was a lucrative innovation because the company also made the glyphosate herbicide—trade-named Roundup—that the GM soybeans could tolerate. This early market success with GMOs convinced Monsanto to purchase several national seed companies, including Asgrow and the DeKalb Genetics Corporation, to gain more direct access to farmers and to high-quality seed germplasm. Monsanto subsequently developed several insect-resistant GM corn varieties that also proved popular with farmers. By 2007, more than 90 percent of all soybean acres in the United States and roughly 60 percent of all corn acres were planted to GM crop varieties that contained at least one of Monsanto's proprietary genetic traits (Etter 2007).

Monsanto jump-started its work on DT corn when it acquired DeKalb, a company with a superb molecular research and breeding hub in Mystic, Connecticut, and excellent hybrid corn genetics. As a further step Monsanto entered into a research exchange agreement with Mendel Biotechnology, a company with a technology it calls Weatherguard, a genetic trait that protects cell walls and membranes during periods of drought and cold. By 2003 Monsanto could announce it had several promising leads on genes for drought tolerance, and in 2004 extensive corn field trials with two different DT genes showed yield improvements ranging from 11 percent to 30 percent under drought conditions (Monsanto 2005). Over the next two years Monsanto then completed fifty-three more large-scale field trials of DT corn, and made public videos showing the leaves of its DT corn plants not rolling up under dryness stress. There is no single agreed standard available to measure stress-tolerance, since the stress can vary from mild to severe, but in 2005 Monsanto's top scientist, Robert Fraley, did tell *USA Today* that "Under severe drought conditions we were able to see 20 percent yield improvement with those plants with the drought gene" (Weise 2005). As of 2007 Monsanto's DT corn technology had moved beyond a simple proof of concept into early product development, and the company announced it would be able to commercialize a DT variety of GM corn for use by farmers in the United States as early as 2010 (Melcer 2004).

Extending Drought-Tolerant Crops to Africa

Monsanto expects that the market for its new DT corn varieties after 2010 will be located primarily in the United States (perhaps 60 million acres), Brazil (30 million acres), and Argentina (6 million acres). The big companies are not developing this new technology for Africa. The DT maize varieties now under commercial development will be yellow maize varieties suited to growing conditions in the temperate zone, rather than the tropical white maize varieties grown by smallholders in

Africa. At a relatively modest cost it would be possible using conventional and marker-assisted breeding to move engineered DT traits into the tropical white maize varieties favored in Africa, but the private companies have little commercial incentive to do this. Most tropical white maize farmers in Africa are still far from becoming valuable seed-buying customers.

Extending agricultural science breakthroughs to the poor is a job that was once, in the 1960s and 1970s, eagerly accepted by donors and private foundations. The falloff in donor support for agricultural science makes this less true today, particularly when the breakthrough in question will be a GM crop likely to encounter high regulatory hurdles, export market anxieties, and alarmist criticisms from NGOs. The same international donors and private foundations that in an earlier day might have jumped at the chance to bring DT crops to Africa are now hesitating to put forward the necessary resources. The scientific challenge is relatively straightforward, yet the financial and institutional efforts needed will be more than routine, and once genetic engineering is involved, daunting regulatory and political barriers will loom. The more a crop improvement initiative plans to use GM crop science, the more skittish key partners and stakeholders will be.

An important early test of donor interest in sponsoring a GM DT crop project for Africa was provided in 2004–05, when Monsanto took the initiative and offered to share its newly discovered DT traits for humanitarian purposes. Monsanto was motivated in equal parts by genuine humanitarian concern, technological pride, and public relations—a desire to counter the continuing effort of some NGOs to demonize the company (they referred to it as "Monsatan"). It was also suspected that if the technology eventually proved a success in the United States, the company could be vulnerable to still more criticism if it had not made an effort to ensure access for the poor. In 2004 Monsanto hired a respected private consultant—Don Doering, previously a senior associate at WRI—to approach donors with the suggestion that Monsanto could be willing to share the new technology, and perhaps even do some of the scientific work of transferring it into tropical crops grown by the poor, if someone else would be willing to pay most of the cost. Doering consulted a significant range of experts and stakeholders in the international scientific and donor communities and then convened an *ad hoc* meeting that brought together representatives from eight different public-sector or nonprofit organizations.

This first meeting led to the formation of an eleven-member international exploratory committee that received enough money from USAID and the Rockefeller Foundation to schedule a larger strategy and planning meeting in 2005. Doering's plan was to use this second meeting to forge a multi-stakeholder consensus and secure donor commitments to form—and finance—a public-private partnership (PPP), with Monsanto playing a partnering but not a leading role. When the larger strategy meeting took place in May 2005 in Arlington, Virginia, it was attended by representatives from both Pioneer and Monsanto, USAID, the Canadian International Development Agency, the Rockefeller Foundation, and CGIAR. Doering presented at this meeting his estimate that getting new GM crops with DT traits into the hands of food-insecure farmers in poor countries would take perhaps a decade, and could cost between $70 million and $150 million (Doering 2005).

The donors that attended this meeting did not open their wallets. They were first of all not comfortable signing on to a plan built around a corporate science breakthrough they did not fully control—or even understand. Pioneer and Monsanto were not yet willing to share details of their research results, either with the donors or with each other, nor were they willing to give up more than limited control over the technology. A royalty-free license for limited use in some countries in some circumstances is what they had in mind, not a freely given donation. The donors also disliked being called together for a pledging session before a fully detailed plan had been prepared.

USAID's misgivings were revealed when it sent only a single technician to the meeting, despite the short taxi ride to Arlington from USAID's downtown Washington, D.C. headquarters. The World Bank (an even shorter taxi ride) sent nobody at all. The International Food Policy Research Institute (IFPRI) sent a single technical scientist to attend the meeting. These Washington-based institutions all had a strong past record of supporting advances in productivity in farm science, yet none was enthusiastic about partnering with Monsanto to bring genetically engineered DT crops to Africa.

USAID's lack of enthusiasm proved critical. Long the leader in supporting GM crops in the developing world, USAID had been working closely with Monsanto on GMOs since 1991, using a core $6 million annual budget for biotechnology that cynics referred to as its "Monsanto earmark." When this DT crops opportunity arose in 2004, however, USAID's biotechnology budget was fully taken up by other projects, leaving the agency with few resources for new science initiatives. USAID had also grown somewhat weary of GMO work in Africa, where hopes for progress had been repeatedly dashed by regulatory burdens and delays. In 2002 USAID had been forced to divert $15 million from its biotechnology program into a new five-year Program for Biosafety Systems (PBS) intended to help poor countries speed up their regulatory approval processes, a diversion that reduced the funds available for new science in areas such as drought tolerance.

The Rockefeller Foundation in 2005 was also a strong traditional supporter of GM crop science, but without a commitment from USAID the foundation alone could not take the lead. Rockefeller's limited resources were also already committed to other projects, including support for a more generic PPP for GM crops in Africa created just two years earlier, the African Agricultural Technology Foundation (AATF), an institution tasked with addressing the corporate patent claims and liability fears Rockefeller had earlier expected would be the greatest constraint to technology uptake in Africa. Rockefeller scientists were sympathetic to the idea of genetically engineered DT crops, and the foundation hinted at one point that it might be able to contribute $1 million a year, but the highly preliminary state of Doering's institutional vision was a sticking point, and Rockefeller specifically wanted a more prominent role for CIMMYT, where considerable progress was already being made on DT maize for Africa using conventional breeding.

CIMMYT had been working to develop DT varieties of both wheat and maize for several decades, making only modest gains with wheat but finding considerable success with maize. For wheat, conventional breeding methods and marker-assisted selection proved to be of limited value (Pellegrineschi 2004). When CIMMYT tried genetic engineering for wheat, it did find a promising lead for drought tolerance with a *DREB1A* gene from *Arabidopsis*, but regulatory restrictions from the

Mexican government prior to 2003 prevented even screenhouse trials from being conducted. When a trial was finally approved and conducted in 2004, CIMMYT learned the DT transgenic wheat performed less well under normal conditions, a disadvantage certain to make it unpopular with farmers (New Agriculturalist 2005).

With maize, CIMMYT's conventional breeding efforts to increase drought tolerance were far more successful, thanks in part to the wider range of natural genetic variation contained within maize plants. Beginning in Mexico in the 1970s and then moving to Africa in the mid-1990s, CIMMYT developed a "stress-breeding" procedure that exposed scores of different varieties of maize to low moisture conditions, then selected those that did best for further development. Systematic field testing of maize crops for drought tolerance is a challenging task that CIMMYT now pursues at 120 separate sites in Africa, including twenty-five sites fully equipped for managed-stress screening. The program operated for years with only a small budget, roughly $3.5 million overall between 1996 and 2004 (Banziger 2007). CIMMYT involved African local farm communities in the research effort through a participatory method called "mother-baby" trials, in which professional researchers and NGOs conduct a mother trial with all the tested varieties in one location at the center of a community, while individual farmers conduct baby trials of subsets of these varieties in their own fields.

Using these low-tech methods, CIMMYT learned it could make steady gains in maize drought tolerance. By 2002 it had been able to develop hybrid maize varieties that gave yields under drought conditions that averaged 20 percent above local hybrids that had not been improved with any stress breeding. Whereas local hybrids averaged 2.53 tons per hectare under drought stress, CIMMYT's hybrids averaged 3.01 tons per hectare, and the best-performing hybrids showed even larger gains (Banziger et al. 2006). Just as important, CIMMYT's DT hybrids experienced no yield loss (in fact a small gain) under normal conditions. These achievements won the King Baudouin Award in 2006, a prestigious CGIAR system research prize.

With such progress being made at low cost through use of conventional breeding methods in Africa, why should the donors want to fund a more expensive approach using genetic engineering? Using genetic engineering remained attractive because it promised to add still greater drought tolerance to the CIMMYT varieties. By using CIMMYT's already improved varieties as a starting point, genetic engineering might be able to offer further step improvements, resulting in an even more dramatic DT advance. The optimal scientific strategy was to use both techniques together, but it was less clear that this would be, in Africa, an optimal political strategy.

Adding transgenic methods to CIMMYT's DT work in Africa would trigger a number of political and regulatory complications. Once the research involved GMOs, separate regulatory approvals would be needed, country by country, for every new greenhouse or field trial. Working through Africa's young, poorly funded, and technically weak national biosafety committees, this approval process would be likely to encounter long and costly delays. Even if no new risks were found, political resistance at the top could still block a final commercial release. Multiple site trials and highly participatory mother-baby trials would be out of the question with GMOs, since Africa's highly precautionary biosafety committees would likely require tight containment and strict segregation from local crops. In Mexico when CIMMYT was finally given official permission to test its DT wheat variety, the regulators had insisted the research be contained within a screenhouse with tightly restricted access. No other wheat plants could be grown within ten meters of the screenhouse trial, and the flowers of the plants had to be covered and isolated from the environment by glassine bags. In addition, the trial had to be monitored by Mexican authorities, and the plant materials had to be destroyed by autoclave at the end of the trial (CIMMYT 2004).

CIMMYT had already seen burdensome regulations slow down some of its other GMO projects in Africa, including a KARI project to introduce insect-resistant GM varieties of *Bt* maize. The project was launched in 2000, but Kenya's National Biosafety Committee did not allow a first set of field trials until 2005. At the same time, a GM cassava project run in Kenya by KARI and the Donald Danforth Plant Science Center had been brought to a standstill for a period when the National Biosafety Committee approved no field trials at all.

Donors at the 2005 meeting in Arlington had other concerns as well.

Even if all the scientific and regulatory hurdles could be overcome, how would GM varieties of DT maize seeds be extended to poor farmers in Africa? Intellectual property concerns were not the issue, since Monsanto had signaled it would be willing, as in some other cases, to share proprietary technologies on a royalty-free basis for humanitarian use in poor countries (Dawe, Robertson, and Unnevehr 2002; SeedQuest 2005). But when private companies agree to this sharing they typically demand a variety of stewardship and quality-control assurances in return, to protect against lawsuits and possible damage to their commercial reputation. Stewardship was a salient concern in the case of maize, a crop that outcrosses so freely through pollen drift that some of the seeds saved for replanting by farmers are likely to be the progeny of plants in neighboring fields. Open-pollinated maize seeds can easily lose their desirable traits within a generation or two. The only certain way to preserve desirable new traits (such as drought tolerance) in maize plants is to purchase new seeds every year from a reputable dealer, ideally a hybrid maize seed dealer with access to the pure parent lines. Yet Africa's poorest maize farmers—the intended beneficiaries of the drought-tolerance initiative—do not usually purchase hybrid maize seed; they continue to plant, year after year, seeds they save from their relatively unimproved open-pollinated varieties (OPVs). This is one reason their productivity is so low.

Technically, it would be advantageous for more poor farmers in Africa to begin using hybrid varieties, since they typically give a yield at least 15 percent higher than even the most improved open-pollinated varieties. Yet many of the local institutions that deliver improved maize seed to farmers in Africa have trouble working with hybrids. Local seed companies can find it difficult to maintain the inbred parent lines needed to scale up seed production, and many NGOs and donors working with the rural poor view the need to repurchase seeds every year as a loss of farm community control. For such reasons CIMMYT has tried to retain the option, when developing its DT maize varieties, of delivering them to companies, donors, and NGOs in an OPV form (Banziger 2007). If the new GM varieties of DT maize were available only as hybrids, to maintain quality control and to satisfy corporate stewardship needs, the task of getting this technology to the poor would be made more difficult. The hybrids-only approach would also become an easy target for NGO critics, who would depict the plan as yet another corporate plot to destroy the traditional independence of small farmers, to make them "dependent" on annual seed purchases from private companies.

Paralyzed by such concerns, the participants at the 2005 Arlington meeting failed to agree on how they could move forward, and Doering's effort to coax public-sector donors into a new DT partnership failed. Yet the idea did not die at Monsanto. In some respects the company was not completely disappointed to see traditional donors such as USAID remaining on the sidelines, since Monsanto knew working with USAID could mean putting up with multiple hassles such as annual funding uncertainties, bureaucratic red tape and delay, sudden changes of direction, and contradictory efforts to please too many diverse political constituencies. The preference of Monsanto scientists was to secure funding for its drought-tolerance project from a less political, less bureaucratic, and more corporate-friendly private foundation—for example, the Bill and Melinda Gates Foundation.

The Bill and Melinda Gates Foundation and DT Crops for Africa

The Bill and Melinda Gates Foundation is still young, created in 1994 when the Seattle-based Microsoft founder Bill Gates and his wife Melinda consolidated their charitable giving by creating a small ($100 million endowment) foundation named for Bill's father. This small precursor foundation focused on two main causes: community needs in the Pacific Northwest and global health. It was transformed into a very big foundation and took its current name in 1999, when it received from Bill and Melinda Gates a massive $16 billion contribution. Four main initiatives were now to be pursued: the Pacific Northwest, global health, education, and libraries. Prior to 2006, the only grants for agriculture given by this foundation were indirect and made through the global health window in the form of various projects to develop food crops with enhanced vitamin, iron, or zinc content, called "biofortified" crops. Lacking a separate agricultural development division, the foundation had no consistent position on GM crops. Although it funded one biofortified sorghum project that planned to use genetic engineering techniques from the start, its $25 million HarvestPlus initiative on crop biofortification, established in 2003 and operated through the CGIAR system, had pledged not to use any rDNA science for at least the first four years (HarvestPlus 2006).

It might have appeared, up to 2005, that the Bill and Melinda Gates Foundation was following the same pattern so many other institutions in Europe and the United States had adopted beginning in the 1980s, a pattern of caring a great deal about Africa yet investing little or nothing in African agriculture or agricultural science. Then in 2005–06, after three years of quiet study, the foundation made a clear move into the agricultural arena. It created a separate agricultural development division and announced as its first large grant a $100 million gift to support a partnership with the Rockefeller Foundation called the Alliance for a Green Revolution in Africa (AGRA), a project based in Nairobi and heavily focused on productivity-enhancing crop science. In this project, as noted earlier, Africans would be trained in plant breeding and then regionally networked to improve local crop varieties. Poor smallholder farmers would be given improved access to the resulting seed improvements through creation of a new Program for Africa's Seed Systems (PASS). When Kofi Annan agreed in 2007 to become the inaugural chair of this AGRA initiative, he explained that the plan at first was not to rely on GM seeds, but nothing was ruled out for the future.

It became clear to Monsanto, watching these developments, that the best way now to pursue a DT crop initiative would be to design something for Africa that the Bill and Melinda Gates Foundation might be willing to fund. Monsanto had reason to expect the foundation would be receptive, since drought tolerance had already been identified by foundation officers as one of the three most important crop-improvement goals—in fact the first of these three—for helping the poor. Gates Foundation staff had visited Monsanto's labs in the spring of 2006 and were so impressed that they hired away the Monsanto vice president who had invited them for the visit, Robert Horsch. Horsch moved to the foundation in November 2006 to become a senior program officer tasked with improving crop yields in sub-Saharan Africa "via the best and most appropriate science and technology, including biotechnology" (Monsanto 2006). As a further sign the Gates Foundation might be receptive to GM drought-tolerant crops for Africa, they had also hired Don Doering.

As a private corporation Monsanto was not eligible to receive funding directly from the Gates Foundation, but it found an eager African partner, the Rockefeller-supported AATF, to take the formal lead. An AATF proposal went to the Bill and Melinda Gates Foundation in 2007, asking for an initial $45 million over five years to develop—in cooperation with CIMMYT—improved varieties of tropical white maize with greater drought tolerance, or what Monsanto now called "yield stability." AATF and Monsanto proposed to take CIMMYT's best DT hybrid maize lines as a starting point, and use the company's lab capacity to improve them through both molecular breeding and genetic engineering to maximize the DT trait. CIMMYT would contribute its white maize germplasm, its expertise in stress breeding, and its local testing infrastructure in sub-Saharan Africa, while Monsanto would contribute high-tech molecular breeding capabilities and also its valuable transgenic biotechnology traits. Monsanto envisioned providing up to four leading genes already identified in its commercial DT crop development efforts, introgressing those genes into CIMMYT's lines, and then waiving its corporate intellectual property rights for use in sub-Saharan Africa. This AATF/Monsanto proposal was designed to offer significant benefits specifically to CIMMYT, in the form of assistance in moving from traditional breeding to marker-assisted breeding. Monsanto's genotyping and fingerprinting labs would be used to provide DNA marker information that would help CIMMYT jump-start its own efforts to advance in molecular breeding.

The benefits to Africa envisioned from this project were described in terms of two different models of yield stability. One new "Yield Stability" hybrid maize variety was envisioned that could provide an additional yield of 1 ton per hectare under moderate drought conditions and an added yield of 0.5 ton per hectare under severe drought, compared to the same hybrid using best-management practices but without the DT trait. Monsanto also modeled a "2× Yield Stability" variety that would provide an added 2 ton per hectare yield under moderate drought and an added 1 ton per hectare yield under severe drought. Monsanto's estimate was that even if only the first yield stability goal were attained, smallholder farmers in Africa currently discouraged by drought threats from using best-management practices, including fertilizer applications, would be given a technological incentive to do so. By one calculation the yield-stability safety net provided by the new seeds might allow farmers to avoid losses or realize a net profit even if hit by moderate drought conditions eight years in a row (Monsanto/AATF 2007). In addition, use of fertilizer and best-management practices in good years would now bring yields high enough to allow farmers to begin diversifying beyond maize production into higher-value and higher-nutrition products, such as beans or vegetables. Monsanto envisioned yield-stable hybrid maize as a safety net in bad years and as a path toward crop diversification and income gains in good years.

AATF and Monsanto knew conventional breeders at CIMMYT had been able to deliver significant yield stability gains without using genetic engineering, and would probably continue to deliver such gains, so they made the case for using GMOs by projecting the longer time it might take to reach a target yield under drought conditions if conventional breeding only were permitted. Using conventional breeding alone at the current annual rate of genetic gain (which they figured was 0.75 percent), it would take forty years to reach the target yield. By adding marker-assisted breeding techniques, the rate of genetic gain would increase to 1.5 percent and the time needed to reach the target would fall to twenty years. By adding genetic engineering, the time to reach the target would fall to thirteen years. AATF and Monsanto were arguing, in effect, that a political decision *not* to use genetic engineering techniques would delay delivery to Africa of the first full DT maize benefits by at least seven years, and of later benefits by even more.

The Monsanto/AATF proposal acknowledged the regulatory complications likely to be faced in Africa once CIMMYT's breeding efforts began to include GM traits, but it expressed a hope that the compelling potential of the new maize would motivate African regulators to take a less skeptical view of the technology. The plan was to begin DT maize development in the Republic of South Africa, where Monsanto already had maize-breeding facilities and where the regulatory environment for research on GMOs was more permissive. Transgenic DT maize field trials might begin in South Africa as early as 2008. Product development would then be completed in Malawi, Kenya, Tanzania, or Uganda, assuming regulators would permit field trials, and if all went well a product could be approved for the market within eight to ten years of the project's onset, assuming consistent future support from the Gates Foundation (Monsanto/AATF 2007).

Would Monsanto scientists be able to deliver on the technical targets contained in this proposal? Their research success with DT maize had so far been confined to the temperate yellow maize varieties they knew best, so it remained to be seen if the transgenes they were working with would perform as well in tropical white maize. As another concern, Monsanto scientists were accustomed to designing products for farmers on quality land with access to the most productive inputs. The company had little experience developing seeds for smallholder African farmers who had acidic soils, sometimes planted too late, didn't use fertilizer, and allowed weeds in the field. Monsanto's plan to rely exclusively on hybrids was another problem. The yield stability offered in the new DT hybrids would give farmers incentives to use not only the improved seeds but fertilizers as well, yet Africa's underdeveloped seed and fertilizer markets might not be able to deliver this package of best-practice technologies to the poor. And would the poorest of the poor be able to afford annual purchases of both new seeds and fertilizers? The solution to this problem might come in the form of seed and fertilizer discounts to the poor, funded by the donor community, but would the same international donor community that had already taken one pass on supporting genetically engineered DT crops in Africa be willing to step in later to subsidize the delivery of such crops to the poor? The political right would criticize the donors for offering subsidies; the left would criticize them for subsidizing a Monsanto-developed GMO technology.

Monsanto's proposal to the Bill and Melinda Gates Foundation built a strong scientific case, but it presented the foundation with a difficult political problem. The Gates Foundation is comfortable with CIMMYT's conventional work on drought-tolerant maize for Africa, and has already awarded a $5 million grant to that project. Funding a new project that relied explicitly on genetic engineering would be a different matter, as it would raise legal, regulatory, and political complications, especially given the involvement of Monsanto. Finding a government in Africa both willing and able to provide the timely biosafety approvals needed to test and eventually commercialize a new GM variety of DT maize remained an uncertainty. Activists would criticize the initiative as a project to promote GMOs driven by a corporate agenda, dependent on hybrids and fertilizers, and hence not tailored to the poor. The promise of greater yield stability under drought conditions might also be dismissed as unimportant relative to the risk of unknown ecological effects. As far back as 1999, Friends of the Earth had included worries about drought-tolerance traits as one of its reasons to call for a complete moratorium on all growing and testing of genetically engineered seeds. Friends of the Earth had worried that DT crops "may have the potential to grow in habitats unavailable" to conventional crops (FoE 1999). How strange that agricultural crops with new growth potential would be seen as a threat by the NGO community, but such was the new political reality.

Science for the Poor, Orphaned by the Wealthy

At a time when so many are trying to do good things for Africa, it is discouraging to see so few willing to help bring the latest in agricultural crop science—crops engineered with enhanced drought tolerance—to Africa's poor farmers. It is African farmers who need this new science the most, yet it seems likely they will be the last to get it.

What has been going wrong? It is tempting to blame USAID and the World Bank, not to mention European donors, for failing to play the traditional public-sector leadership role. Yet significant culpability in this case also attaches to government leaders in Africa. One source of USAID's reluctance to offer funding for a GM DT crop initiative was a lack of enthusiastic governmental partners at the African end. USAID had grown tired of promoting a technology that governments in Africa apparently did not want. Government leaders in Africa had repeatedly endorsed agricultural biotechnology in the abstract, and some had even been willing to approve greenhouse and field-trial research (as long as donors paid for the greenhouse), but when it came to approving the release of GMOs for commercial use, these same African elites repeatedly got cold feet. Even a GM trait as attractive as drought tolerance made African governments nervous, simply because it was a GM trait. Volunteering to be the entry point into Africa for a new GM crop—especially one developed by Monsanto—was a role few African political leaders found attractive. So they held back, taking as their excuse the fact that the performance of the technology in Africa—and the donor funding—both remained uncertain. It was telling that there were no African government partners present at the May 2005 Arlington meeting.

At the technical level, African scientists and policy analysts are fully aware of the importance of bringing modern biotechnology to the continent. In April 2007 a High-Level African Panel on Modern Biotechnology detailed the contribution biotechnology could make to food security, nutrition, health care, and environmental sustainability in Africa. Yet the co-chairs of this panel, Calestous Juma and Ismail Serageldin, emphasized that expert endorsements would not suffice. More support was needed from donors, and more courage and firmness was also needed from African heads of state and governments (Juma and Serageldin 2007). Three years had passed since Kofi Annan placed his own challenge in front of Africa's heads of state in Addis, with a call for a Green Revolution in Africa. A new commitment of public-sector money from donors would go nowhere without governmental buy-in from Africa. Going around the public sector, hoping for a possible partnership between AATF, Monsanto, and the Gates Foundation, is an untested, second-best approach. Even if the Gates Foundation makes a full funding commitment to the Monsanto/AATF DT maize project for Africa, and even if Monsanto and CIMMYT scientists are able to overcome all the scientific and technical challenges, Africa's response to the technology at the political and regulatory level will remain uncertain.

If this worthy initiative is slowed or blocked by a lack of political enthusiasm at the African end, critics of GM crops will look for a way to blame Monsanto and its "silver bullet" crop scientists. Yet the real failure will have been a lack of public-sector leadership and vision. Instead of investing adequately in modern crop science to help Africa's poor, public-sector institutions, both foreign and African, have continued to discount the value of agricultural science and have continued to underfund or stifle the science behind GMOs. Without a more science-friendly public-sector posture, even the most generous private philanthropists working with the most capable corporate scientists will not be able to deliver needed benefits to Africa's poor.

CONCLUSION

An Imperialism of Rich Tastes

In a world badly divided between rich and poor, the policy preferences and regulatory standards of the rich tend to prevail, sometimes to the disadvantage of the poor. In Chapter 1 of this book I noted that many citizens in wealthy countries are skeptical toward genetically engineered crops, in part because they know they can remain rich and well fed without them. In Chapter 2 I discussed this as part of a larger pattern of prosperous countries turning away from all new forms of agricultural science, because they have seen too much of their own farming tradition altered by agricultural science and because they don't need any more. As stated in Chapter 3, this turn against modern farm science is now being exported to the poor countries in Africa, even though agriculture there is burdened by having too little science rather than too much. Despite Africa's needs, international assistance to agriculture there has been cut back sharply since the 1980s, and many NGOs from wealthier countries are now warning Africans away from the science-based path to agricultural productivity, even though this was the path out of poverty earlier used by the rich.

Turning from agricultural science in general back to GMOs in particular, in Chapter 4 I examined the notion that the European countries with greatest postcolonial influence in Africa have used their foreign assistance programs, their influence over the UN system, commodity markets, and NGOs to persuade governments in Africa to adopt a highly precautionary regulatory stance toward GM crops. This European-style regulatory approach has so far prevented any farmers in Africa—outside of the Republic of South Africa—from growing any GM crops. Regulations that block the use of GMOs may be affordable for Europe, where farmers are already highly productive and consumers well fed, but this approach imposes a damaging constraint on farm productivity in Africa. This extreme regulatory standard is imposed for no good reason, given the accumulating evidence that GM crops are no more dangerous to human health and the environment than conventional crops. I then examined in Chapter 5 initial political responses to a new application of genetic engineering in agriculture, the insertion of drought-tolerance genes into farm crops. Even though crops better able to maintain yield under drought conditions would provide obvious benefits for Africa's rural poor, the first political reaction to this new technology from the traditional donor community has been hesitant. Some donors were no longer investing significant efforts or energies in any kind of agricultural science; others were not interested in promoting this new drought-tolerance breakthrough because it would mean working with GMOs.

In this concluding chapter I finally ask why so many political leaders in Africa have been willing to follow the lead of the rich, particularly the European rich, in cutting back on public-sector investments in agricultural science and placing stifling regulations on agricultural GMOs. Why have these policy tastes of the rich been adopted by governments in Africa, where poverty remains pervasive and where more rather than less agricultural science is so desperately needed? In this book so far the dominant pattern of policy choice has been one driven by material interest—for example the choice of the rich to invest more in medical science and less in agricultural science, parallel to the much larger material need they now feel for new medicines versus new farming techniques. So why have African governments cut back on their own agricultural science investments at a time when their material needs in the farm sector are not being met, with their faltering agricultural systems in such obvious need of technical improvements? Only because they had no material interest in agricultural GMOs did Europeans decide to block this new technology with stifling regulations; why are governments in Africa doing the same thing, despite the promise of GMOs as a new source of productivity to the two-thirds of their citizens who are farmers?

Africa's embrace of a European-style regulatory approach to GMOs is doubly interesting because it is the opposite of what many early critics of globalization might have expected. Their prediction had been a "race to the bottom" in which the rich would be forced to accept the regulatory standards of the poor, not the other way around. As global capital became more mobile, private companies would be expected to move their investments to countries with less-demanding regulatory systems, for example to countries with lower labor and environmental standards. Then in an effort to compete, wealthier countries would be obliged to lower their regulatory standards as well, and a race to the bottom would ensue, eventually lowering regulatory protections everywhere. The opposite is happening in the case of agricultural science and agricultural GMOs: African countries have been adopting the regulatory standards of the rich.

Trading Up

Africa's behavior in this case should not be entirely surprising. Recent studies of regulatory standards under globalization emphasize the possibility of an upward convergence of standards, a slow drift to the top rather than a race to the bottom or even a convergence at the middle. A race to the bottom was always implausible because when governments make regulatory standards they pay attention to much more than corporate preferences; they also consider the preferences of voters and their elected representatives, powerful interest groups such as labor unions or consumer and environmental advocates, and the preferences of regulators themselves inside the government. Moreover, when private firms make investment decisions they pay attention to much more than regulatory costs; they also consider national market size, quality of institutions and infrastructure, and skill level in the workforce. This is why roughly three-quarters of all foreign direct investment today goes not to countries with low standards but to countries with high standards, the rich OECD countries where labor, environmental, and consumer protections are highest. Daniel Drezner's work shows that within the OECD region labor standards have converged upward rather than downward, and beyond the region a slow upward drift in the enforcement of core labor standards has occurred as well. Environmental standards have behaved similarly, with an upward convergence occurring within the OECD region in addition to slow and erratic upward movements toward stronger environmental protection in the developing world (Drezner 2001).

The most frequently cited explanation for upward convergence in regulatory standards is a mechanism that works through international trade, a mechanism David Vogel labeled "trading up" (Vogel 1995). Private commerce drives everyone's standards up because the wealthiest countries (with the highest standards) are also the world's largest importers of goods and services. Foreign suppliers that want to sell to the rich must produce the goods the rich demand, at the quality level they demand, and even with the production processes they demand. This implies honoring the regulatory standards—both the product standards and production process standards—favored by the rich countries.

The European Union is now an economic bloc of twenty-seven nations containing more than 480 million mostly affluent consumers. It has strict product safety, consumer protection, environmental, and health regulations—all laid out in a growing body of law currently 95,000 pages in length (Buck 2007). Whenever the European Union tightens the product-safety and production-process standards it now imposes on its own domestic firms, private companies abroad seeking to maintain access to the European market must adjust to this new standard as well. The European Union is the big customer, and the customer is always right. Because of the importance of its import market, Europe's high regulatory standards are gradually becoming something like a new global norm. In 2003 a *Wall Street Journal* article complained, "Americans may not realize it, but rules governing the food they eat, the software they use and the cars they drive increasingly are being set in Brussels" (Mitchner 2003).

Private firms even in the United States can find it inconvenient to have to adjust to European rules. In a 2005 study titled "Exporting Precaution," Lawrence A. Kogan argued that Europe's pursuit of a "risk-free regulatory agenda" had become, through the trading-up mechanism, a significant new burden on business expansion options in the United States (Kogan 2005). In a 2006 review of international policies for chemicals, electronic wastes, and hazardous substances, Henrik

Selin and Stacy VanDeveer showed that "the EU is increasingly replacing the United States as the de facto setter of global product standards" (Selin and VanDeveer 2006, p. 14). In 2007 C. Boyden Gray, the U.S. Ambassador to the European Union, accused Brussels of seeking to "export their regulations abroad" in ways that would drive up global production costs and harm U.S. commercial interests (Buck 2007).

In Chapter 4 of this study we saw that the United States has indeed been forced to adjust to Europe's highly precautionary regulation of GMOs. Informally, American regulators and private industry have decided to keep several new GM crop varieties off the market at home—including GM wheat and GM rice—so as not to risk a loss of export sales to regions where these crops are unapproved, including Europe and Japan. American farmers had learned their lesson in 1998 when Europe stopped buying bulk shipments of corn containing some varieties that regulators in Brussels had not yet approved. In Africa, GMO policy has been influenced even more by the trading-up mechanism. Because African states depend so much more on Europe as an export market, the best way for them to avoid losing sales is to adopt Europe's regulatory approach outright, imposing strict precautions on GMOs even if this means (as in Europe) that the technology will go essentially unused. Some African governments have even been willing to reject GMOs as food aid in a drought emergency, so as to remain GMO-free and ensure their commercial access to European markets. In Zambia we saw that a private company selling organic baby corn to Europe helped lobby the government—successfully, despite a drought emergency—to impose a total ban on imports of GM yellow maize as food aid from the United States.

Scholars of globalization will have noticed that several known mechanisms of upward convergence beyond trading up are not operating in this case. Sometimes multinational firms lobby host governments in poor countries for higher standards, because they sense such standards will be easy for them to meet while imposing a burden on less-capable local competitors (Graham 2000). This sort of corporate lobbying is not in evidence here, since the regulatory standards being set in Africa are far more demanding than even the most capable private firm would desire. Also missing in this case is the role sometimes played by transnational networks of scientific and technical experts, given the name "epistemic communities" by the political scientist Peter Haas (Haas 1992). According to Haas, when governmental leaders are uncertain about the implications of a policy choice, expert opinions by scientists can be decisive in driving them to adopt higher regulatory standards, especially in the area of environmental protection. However, in the case of agricultural GMOs, the consensus view of scientists has largely been ignored by policy makers in Europe and also by the African governments now following Europe's lead. The consensus view among scientific authorities is that none of the GM foods or crops commercialized so far has created any new risk to human health or to the environment, yet in Europe—and Africa—the noose of governmental regulation continues to tighten around the technology.

Yet there is more than just trading up going on here. In Chapter 4, I demonstrated that Africa's upward convergence toward European-style regulatory standards was also being driven by European foreign assistance policies, European influence within the UN system, and European NGO activism. Following a brief summary of each of these forces, I will go on to examine why they have had such a strong impact on political elites in Africa.

Foreign Assistance Policies

It is common to dismiss state-to-state foreign aid to poor countries either as an antiquated tool from the Cold War era, or an application of early development theory that never quite delivered on its promises, yet in this study we found that in Africa foreign aid is still an important channel of external influence. For cash-poor governments in Africa, new investments in agriculture and agricultural science are usually not made unless financed through foreign assistance. The importance of this external support was revealed when national government spending in Africa stopped growing in the 1990s after external support was cut back. In the 1980s and 1990s the share of U.S. foreign assistance that went to agricultural development fell from 25 percent to just 1 percent. The share of bilateral European aid going to farming in poor countries fell by roughly two-thirds. The share of World Bank lending that went to the agricultural sector fell from 30 percent to just 8 percent. This had little impact on the more advanced developing nations of Asia and Latin America, who no longer needed external assistance to boost the productivity of their farming sectors, but it did have a large impact in Africa, where the smallholder farming sector had not yet taken off and where dependence on foreign aid remained acute. When external assistance to African agriculture was cut—by more than 40 percent during the decade of the 1990s alone—Africa's own agricultural budgets also had to be cut, and poor farmers there fell further behind.

Foreign aid critics try to dismiss those—from Jeffrey Sachs to Tony Blair—who have urged directing greater external financial resources toward Africa to give the continent the "big push" it needs to begin moving out of poverty. The economist William Easterly has made a career out of publishing books that remind us of what foreign aid *cannot* do from the outside to save Africa (Easterly 2007). This is a dismissal that ignores the many good things foreign assistance has done in Africa in areas such as public health and education. Externally funded projects in these areas helped bring sharp reductions in both infant mortality and illiteracy. The weakening of agricultural science efforts in Africa due to external assistance cutbacks in the 1980s and 1990s is a sad negative example of this capacity for aid to make a difference.

Donor assistance to Africa overall has been increasing as never before, yet aid to farming—particularly if it involves bringing in western science—has curiously dropped off the agenda. Not since the 1970s, when Norman Borlaug won a Nobel Peace Prize for having helped develop the new seeds that ended famine threats in India has agricultural science enjoyed respect in wealthy nations. More recently, from their own position of postindustrial comfort, the rich have persuaded themselves that the Green Revolution could not possibly have been good in Asia, either for the poor or for the environment. They have moved away from advocating modern agricultural science as an appropriate response to Africa's farm productivity crisis, thinking it better for Africans to keep western science at a distance and seek solutions based on their own internal resources, relying on the indigenous knowledge and cultural strength of their own local communities and their own existing storehouse of traditional crop varieties. This turn against agricultural science has somehow managed to endure despite a continual worsening of agricultural circumstances in Africa. When the Norwegian Nobel Committee awarded the coveted Peace Prize to an African in 2004, it selected not a scientist but a Kenyan environmental activist named Wangari Maathai, who in her Nobel lecture stressed the importance of community-based self-help using "indigenous seeds and medicinal plants" (Maathai 2004).

Taste makers in the donor community are forgetting the vital role agricultural science played in their own material success, when it released labor from farming and simultaneously brought a more abundant supply of food to the market. If serious about helping Africa, the rich should be investing in new agricultural science to bring a similar increase in productivity on African farms. Instead, citizens in rich countries are now preaching to Africans the merits of organic farming, which defines itself in terms of the applications of modern science (synthetic chemistry and genetic engineering) that it will *not* allow farmers to use. Organic farming is an expensive option, and in even the richest countries it has spread to provide only a small fraction of total farm production. In Europe organic products are grown on only 4 percent of total farmland, and in the United States on just four-tenths of one percent of farmland. Nonetheless, rich-country donors and NGOs are telling Africans they should forswear using nitrogen fertilizers in favor of more laborious and less productive organic production methods.

Rather than assisting Africa in the development and delivery of productive farming technologies, wealthy nations and NGOs are assisting Africans in keeping such technologies at a distance, likening GMOs to "contamination" and regulating biotechnology like a hazardous waste. The GMOs currently on the market have helped farmers become more productive while doing no new documented harm either to human health or the environment, yet European donors and NGOs are working to keep this technology out of Africa. Instead of helping Africans become more productive or better fed, these donors and NGOs are telling them to become more vigilant: still poor and still hungry, but at least "safe" from GMOs. Foreign assistance should be used to give Africans help in addressing the real and documented risks to food safety and biodiversity that confront the continent, such as lack of refrigeration, or destruction of wildlife habitat by ever-expanding, low-yield agriculture. External donors and NGOs pay little attention to these genuinely African problems, preferring instead to export their own concerns from back home in Europe or America.

Campaigns by NGOs

Nongovernmental organizations from prosperous countries deserve full praise for having advanced essential human values in poor countries, including public health, education, human rights, and gender equity. With courage and great energy, NGOs bring life-saving humanitarian assistance to vulnerable communities when governments fail or fail to act. In the case of agricultural science, however, the recent record of many groups in global civil society has been harder to defend. NGOs from rich countries too often depict modern agricultural science as the problem rather than the solution. Having seen what they view as an excess of science-based farming back home in Europe and North America, they make the error of warning Africa away from modern agricultural science altogether.

For farmers in Africa today, productivity is low and poverty high because far too little science has been brought to farming. Currently only 4 percent of Africa's farmland is irrigated, less than 30 percent is planted to improved seeds, and average fertilizer use is only 9 kg per hectare, compared to 117 kg per hectare in the industrial world. Excessive fertilizer use in farming remains a problem to worry about in wealthy countries, but in Africa the problem is too little nitrogen in the ground rather than too much. Factory farming and corporate concentration in agriculture are appropriate targets for NGO action back home in Europe and North America, but the danger in Africa's impoverished countryside is that private companies with modern technologies will invest too little rather than too much.

For the NGO community, opposition to GMO crops has emerged as part of a larger opposition to all science-based productivity growth in farming. GMOs have simply replaced the Green Revolution as the most frequently cited illustration of this supposed peril. In 2004 more than 600 NGOs sent a letter to the director general of FAO describing GMO crops as dangerous because they would only replicate the "tragedies" inflicted on the poor by the original Green Revolution. Avoiding famine and reducing poverty was a tragedy? This NGO rejection of farm science no longer commands an audience in Asia, where the broad material benefits of several decades of technological innovation are now visible for all to see, but in Africa this critique can still give decision makers pause.

It is telling that when NGOs from rich countries go to work in Africa, they do not oppose modern science across the board. They have often been among the first to demand access for Africans to modern medical science (such as drugs to treat HIV/AIDS) and to modern information technologies (such as computers and cell phones to close the "digital divide"). Since modern drugs and new computers are valued in Europe and North America, NGOs are comfortable promoting both in Africa. Modern agricultural science is no longer valued back home, so NGOs are comfortable trying to keep it out of Africa.

The dismissive view of modern agricultural science brought into Africa by some NGOs has been perversely attractive to urban policy elites in Africa. It pleases these elites to be told they should continue staying away from GMOs, continue underutilizing chemical fertilizers, and continue underinvesting in modern crop science, relying instead on traditional crop varieties and indigenous knowledge. Such advice is easy to accept since it means these African leaders don't have to change any of their current policies; they can continue ignoring agriculture.

Even as NGOs warn Africa away from western science in the farming sector, they eagerly promote an extension of western regulatory standards into Africa. Importing the latest farming technologies from the West is bad, but importing the latest in highly precautionary biosafety regulation is good. Growing more regulations in Africa now gets higher priority than growing more food. It is not an influx of unregulated technologies that is harming rural Africa today, but instead the almost complete nonavailability of new technology.

The United Nations

Another external mechanism influencing Africa to adopt an overly restrictive approach to agricultural biotechnology has been the United

Nations Environment Programme (UNEP), which has encouraged governments throughout the developing world to adopt National Biosafety Frameworks based on European-style precaution as part of their adherence to the 2000 Cartagena Protocol. This UNEP program has had only modest influence in shaping national policies outside of Africa because dependence on foreign aid there is much lower, national regulatory capacity tends to be much higher, and because these other countries already were on the way to developing their own systems for regulating biosafety, making the transplant of an entirely new system by UNEP more difficult. In Africa, which was more of a blank slate for biosafety regulation, UNEP's activities have been influential.

The decision of other international institutions to duck the controversial GMO issue helped give UNEP more room to operate. The traditionally pro-science organizations that might have been expected to promote GMOs in Africa—such as FAO, the World Bank, and the CGIAR—mostly held back or backed off, so as not to jeopardize European support and funding. Had these pro-science institutions played a stronger role, Africans working inside agricultural ministries, university faculties, and national agricultural research systems would have had the arguments and the resources to defend the new technology and fight off the strange claim that GMO crops should be regulated under the Cartagena Protocol in the same manner as hazardous chemical wastes.

The UNEP/GEF program had a much stronger influence in Africa than elsewhere because in most cases there were no existing regulations on the books needing to be changed. Prompted by UNEP, one African nation after another has gone from having no GMO biosafety regulations at all to embracing a regulatory approach just as precautionary as in Europe. All but one of the twenty-three African countries that completed their National Biosafety Frameworks under the UNEP/GEF program by 2006 started out as "blank slate" countries with no previous laws on the books governing GMOs, and twenty-one of the twenty-three in the end selected the strongest possible regulatory posture—what UNEP calls a Level One approach—requiring the equivalent of an act of parliament before a commercial release of the technology can be considered. It is no accident that the only one of these twenty-three countries so far to have approved the planting of any GM crops by farmers is the Republic of South Africa, which had its own functioning biosafety system in place before the UNEP/GEF effort was launched. In all the rest where National Biosafety Frameworks have been completed, farmers are not yet allowed to plant any GMOs.

Why Follow Europe Rather than America?

It is sometimes assumed that globalization will bring a *de facto* Americanization of standards and tastes. Not in this case. The external influences that have shaped policies toward GMOs in Africa have come primarily from Europe rather than the United States. This reflects Europe's much greater proximity to Africa's policy elites commercially, financially, geographically, historically, and culturally as well.

Commercially, European markets are far more important to African exporters than the American market. The European Union countries buy roughly €7 billion in farm products from Africa every year—six times as much as the United States buys. Through the mechanism of trading up, Africa must imitate European product and production standards. Financially, official development assistance from Europe to sub-Saharan Africa is roughly three times as great as assistance from the United States. Through international financial institutions as well, it is primarily European money that now reaches Africa. Recall that the collapse of American assistance to international agricultural science made Europe by itself the source of 41 percent of funding to the CGIAR system. It is thus unsurprising to see the system directing only 3 percent of its budget to work on GMOs. Also, since Europe now contributes three times as much as the United States to the GEF Trust Fund it is unsurprising to find UNEP using GEF money to promote European-style precaution toward GMOs rather than American-style permissiveness.

Europe's closer historic and cultural ties to Africa are also a powerful influence. Scholars of regulatory harmonization notice that a shared history, language, or religion can create "psychological proximity" among nations otherwise geographically or economically distant from each other (Rose 1993). Psychological proximity to Europe is palpable today among the urbanized elites who dominate governments in postcolonial Africa. They converse in European languages, follow European religious practices, read European newspapers, play European sports, send their children to European universities, deposit their money in European banks, and travel to Europe as often as possible for holidays and shopping. Sometimes these Africans will confide to visiting Americans that they do not always feel warmly welcomed in Europe, but still it is to Europe that they normally defer. If a visiting American suggests the possibility that GMOs might actually be safe to eat, the first African response will usually be, "Oh? Then tell me why nobody in Europe eats GMOs." Africa's recent aversion to GMOs is at some basic cultural level an emulation of Europe's aversion to GMOs.

Not every developing region follows the European lead so closely. In Latin America, the geopolitical backyard of the United States, local policy elites have been far more likely to follow America's lead. As of 2006, only twelve developing countries around the world had approved the commercial planting of any GMO agricultural crops, and seven of these twelve were Latin American countries: Argentina, Brazil, Paraguay, Uruguay, Mexico, Colombia, and Honduras. It is telling as well that the only Asian developing country in 2006 to have approved the commercial planting of a GM crop other than cotton was America's former colony in the region, the Philippines.

Internationally Reinforced Urban Bias

European regulatory standards toward GM crops are thus spreading through Africa thanks to a powerful combination of international mechanisms including commodity trade, foreign aid, NGO campaigns, the activities of UNEP, and postcolonial emulation. The confluence of these mechanisms bears a strong resemblance to the "structural theory of imperialism" originally offered in 1971 by Johan Galtung, a theory based on close psychological proximity between urbanized elites in poor countries and counterpart elites in rich countries (Galtung 1971). Galtung argued that national policy makers in the capital cities of peripheral countries can actually come to feel a closer harmony of interests with policy makers in the capital cities of rich countries than with the poor people living in the "periphery" of their own country. In Galtung's terminology, these elites in poor countries at the center of the periphery feel closer to elites at the center of the center than they do to the majority of their own citizens, living in the periphery of the periphery. The mechanisms originally referenced by Galtung in explaining this imperial capture of elites in poor countries included several of those stressed here: export dependency, financial dependency, and cultural affinity.

To suggest that urbanized policy elites in Africa might feel closer to Europeans than to their own rural poor is a harsh charge, and not an immediately convincing one since even the most comfortable members of Africa's urban political class make regular return visits to "their village" so as to maintain family and ethnic ties and display generosity at weddings or funerals. When Galtung was writing in 1971 he had in mind the urban political elites of Latin America more than those of Africa. Yet the specific solidarity urbanized elites in Africa continue to feel toward their own village, family, and ethnicity is seldom given larger expression in policies of generalized help to all the rural poor. Recall that in Africa most governments now devote only 5 percent of their budget to the agricultural sector, even though up to two-thirds of their citizens, and 80 percent of their poor citizens, still depend on this sector for income and employment.

Although African societies are still primarily rural and agricultural, Africa's institutions and instruments of political power have been highly urbanized since colonial days. When the colonizing Europeans came to Africa, they created new institutions of taxation and governance conveniently centralized in the capital city, which was itself often a new creation located on the coast to make coming and going from Europe less difficult. When independence from colonial rule arrived in the 1960s, the departing Europeans handed off these centralized urban instruments of governance to a new local elite of equally urbanized, European-educated Africans, who began operating them in much the same manner as the former colonizers. Africa's new governments imposed heavy tax burdens on farming just as before, and just as before most job and service benefits were directed toward urban dwellers. To the present day governmental service delivery in Africa is more urbanized than anywhere else in the world, even though Africa itself is the least urbanized of continents. In one study of nineteen developing countries, including six countries from sub-Saharan Africa, four of the five that were most urbanized in terms of public service delivery were African (Nigeria, Cote d'Ivoire, Burkina Faso, and Senegal), whereas the seven most decentralized were all non-African (Tuskegee University 2001).

With all the criticisms recently leveled against poor governance in Africa, it is surprising how few have been directed against this excessive urban bias. The conventional measures of governance used by the World Bank ignore the urban-rural divide, focusing instead on whether governments are democratic or not, corrupt or not, and subservient or not to the courts. Such measures show the continent to be highly diverse in its performance, with some states—such as Tanzania, Liberia, Rwanda, Ghana, and Niger—having made substantial progress over the past decade (Kaufmann, Kraay, and Mastruzzi 2007). For Africa's poor farmers living in the countryside, however, what matters most isn't these various *traits* of government but instead the *actions*—or inactions—of government. Are sufficient public investments being made to build and repair rural roads, rural clinics, and rural schools, and to develop and extend new farming technologies? Even in countries in Africa where elections are now free and fair, where officials are relatively honest, and where rule of law now tends to prevail, public investment in the creation of these rural public goods remains weak.

Africa's political class is so centralized and urbanized that foreign visitors commonly find trips out of the capital city will be considered unnecessary by their hosts. Everyone you might need to see is right there, usually only a short distance from the international airport and your centrally located downtown hotel. Even if your business is agriculture, all of the key government officials, university scholars, think-tank researchers, farmers' union leaders, international donors, foundations, and NGOs you may wish to visit are close at hand in the same single downtown space. A trip out of the capital may be recommended for the weekend, but most likely this will be by plane to a nature park rather than by road to visit farms, since auto travel into the countryside is still taxing, logistically unpredictable, and sometimes physically unsafe. Between downtown appointments the visitor will observe enough urban poverty through the window of a taxi to imagine that conditions in the villages probably are not much worse. Yet it is precisely because poverty is much worse in the countryside that Africa's urban slums are filling so fast.

Comparisons between urban and rural poverty in Africa are conveniently obscured by a scarcity of easily accessible data. The World Bank every year publishes a compilation of more than 900 world development indicators presented country by country and region by region in over eighty separate tables, yet rural poverty measures are almost nowhere to be found in these tables. Nineteen of these eighty tables are dedicated to measuring the circumstances of people, yet in only one of these nineteen tables do we even find a distinction drawn between people living in cities and towns versus those living in the countryside. This table attempts to offer an urban versus rural poverty breakout, with columns that show the percentage of urban versus rural citizens falling below national poverty lines. Unfortunately, for more than half of the sub-Saharan African countries listed in this table, the columns are empty (World Bank 2006a). Where data are available, they show the share of the urban population below the poverty line averaged 36 percent in Africa, compared to a poverty rate among the rural population of 56 percent. The impoverishment of Africa's countryside is also obscured in the way the World Bank has been monitoring progress toward a new set of development objectives—the so-called Millennium Development Goals. All eight of the Millennium Development Goals, and seventeen out of the eighteen specific targets under the goals, ignore urban versus rural differences completely. Only in tracking the tenth target, to halve by 2015 the proportion of people without sustainable access to safe drinking water and basic sanitation, is the World Bank using a set of monitoring indicators that distinguish between urban versus rural conditions (World Bank 2006a).

In our rapidly urbanizing world, the rural poor in Africa are rudely being ignored. They work long and hard at farming, but continue to make little gain because they use seeds, animals, tools, and other inputs that are essentially unimproved by modern science. Compared to small farmers in other regions who have now experienced enough of a technology upgrade to see their productivity and incomes rise, smallholder farmers in Africa remain starved for science.

The investments now needed to bring useful applications of modern science to African farming—including modern crop science, animal science, and chemical science—are not likely to come spontaneously from Africa's own highly urbanized policy elites, who have only meager resources to invest and who are distracted by too many other priorities. Private companies lack a commercial incentive to do the job. Africa's agricultural science deficit will thus have to be corrected through supportive interventions from international donors. The donor community briefly provided the external assistance African farming needed in the 1960s and 1970s, but then in the 1980s and 1990s when the support of their own citizens for agricultural science collapsed back home, the rich donors pulled away from extending technical support to agriculture in Africa as well. Restoring this external support should now be an urgent priority.

In the specific case of biotechnology, rich countries are imposing their tastes on Africa through more than just a lack of donor support. Through multiple channels of direct and indirect influence, urban elites especially from Europe are persuading counterpart urban elites in Africa to hold GMOs in suspicion and do without them, or place them under stifling regulatory control. This is an acceptable posture in Europe, where farmers make up only a tiny share of all citizens and can remain productive and prosperous without GMOs, but it is highly inappropriate to the needs of rural Africa. Europeans are imposing the richest of tastes on the poorest of people.

So in the end it is not the citizens of Africa who are rejecting agricultural biotechnology. The technology is being kept out of Africa by a careless and distracted political leadership class that pays closer attention to urban interests and to inducements from outsiders—from European donors, UN technical advisors, NGOs, and export market customers—than to the needs of their own rural poor.

REFERENCES

Adams, Alayne M., Jindra Cekan, and Rainer Sauerborn (1998). "Toward a Conceptual Framework of Household Coping: Reflections from Rural West Africa," *Africa: Journal of the International African Institute,* 68 (2): 263–283.

Africa Center for Biosafety (ACB) (2004). "GM Food Aid: Africa Denied Choice Again?" Africa Center for Biosafety, Earthlife Africa, Friends of the Earth Nigeria, GRAIN, and SafeAge. May.

African Biotechnology Stakeholders Forum (ABSF) (2006). "Biotech Status in Africa," www.absfafrica.org.

AgraFood Biotech (2006). "Report: Biopharmaceuticals to Reach $12 Billion by 2012," 172, (13 March): 7.

Akst, Daniel (2003). "Cheap Eats," *Wilson Quarterly,* 27 (3, Summer): 30–41.

Alston, Julian M., Philip G. Pardey, and Vincent H. Smith (1998). "Financing Agricultural R&D in Rich Countries: What's Happening and Why," *Australian Journal of Agricultural and Resource Economics,* 42(1): 51–82.

Alston, Julian M., Daniel A. Sumner, and Stephen A. Vosti (2006). "Are Agricultural Policies Making Us Fat? Likely Links between Agricultural Policies and Human Nutrition and Obesity, and Their Policy Implications," *Review of Agricultural Economics,* 28 (3, Fall): 313–322.

Alston, Julian M., Steven Dehmer, and Philip G. Pardey (2006). "International Initiatives in Agricultural R&D: The Changing Fortunes of the CGIAR," pp. 313–360 in Philip G. Pardey, Julian M. Alston, and Roley R. Piggott, *Agricultural R&D in the Developing World: Too Little, Too Late?* Washington, DC: IFPRI.

American Soybean Association (ASA)(2006). "American Soybean Association Calls EU Traceability and Labeling Review a Whitewash," *SeedQuest* News Release, St. Louis, 10 May.

Anderson, Kym (2006). "Reducing Distortions to Agricultural Incentives: Progress, Pitfalls, Prospects," *American Journal of Agricultural Economics*, 88 (5): 1135–46.

Angell, Marcia (2004). *The Truth About the Drug Companies*. New York: Random House.

Annan, Kofi (2004). "Africa's Green Revolution: A Call to Action," Remarks at "Innovative Approaches to Meeting the Hunger Millennium Development Goals in Africa," Addis Ababa, 5 July. Press Release SG/SM/9405 AFR/988.

Arts, Bas, and Sandra Mack (2007). "NGO Strategies and Influence in the Biosafety Arena, 1992–2005," pp. 48–64 in Robert Faulkner, ed., *International Politics of Genetically Modified Food*. New York: Palgrave Macmillan.

Banziger, Marianne (2007). Director, Global Maize Program, CIMMYT. Personal interview, Nairobi, May.

Banziger, Marianne, Peter S. Setimela, David Hodson, and Bindiganavile Vivek (2006). "Breeding for Improved Abiotic Stress Tolerance in Maize Adapted to Southern Africa," *Agricultural Water Management*, 80: 212–224.

Bauer, Martin W. (2005). "Distinguishing Red and Green Biotechnology: Cultivation Effects of the Elite Press," *International Journal of Public Opinion Research* 17 (1): 63–89.

Beck, Ulrich (1992). *Risk Society: Towards a New Modernity*. New Delhi: Sage.

Beckfield, Jason (2003). "Inequality in the World Polity: The Structure of International Organization," *American Sociological Review*, 68 (3, June): 401–424.

Beintema, Nienke M., and Gert-Jan Stads (2004). "Investing in Sub-Saharan African Agricultural Research: Recent Trends," International Food Policy Research Institute, 2020 Africa Conference Brief 8. Washington, DC: IFPRI.

Belasco, Warren (1989). *Appetite for Change: How the Counterculture Took on the Food Industry, 1966–1988*. New York: Pantheon.

Bennett, Jon (2003). "Food Aid Logistics and the Southern Africa Emergency," *FMReview* 18 (September), http://fmreview.org/text/FMR/18/12.htm.

Benson, Richard (2006). *The Farm: The Story of One Family and the English Countryside*. London: Penguin Books Ltd.

Bernauer, Thomas, and Erika Meins (2003). "Technological Revolution Meets Policy and the Market: Explaining Cross-National Differences in Agricultural Biotechnology Regulation," *European Journal of Political Research*, 42 (5): 643–683.

Bill and Melinda Gates Foundation (2006). "Bill & Melinda Gates, Rockefeller Foundations, Form Alliance to Help Spur 'Green Revolution' in Africa," 12 September. Seattle: Bill and Melinda Gates Foundation.

Biotechnology and Development Monitor (1994). "Editorial: Stress Seldom Comes Alone," *Biotechnology and Development Monitor*, 18 (March): 2.

Block, Steven A. (1994). "A New View of Agricultural Productivity in Sub-

Saharan Africa," *American Journal of Agricultural Economics*, 76 (August): 619–624.

Bocker, Andreas, and Giuseppe Nocella (2005). "Trust in Authorities Monitoring the Distribution of Genetically Modified Foods: Dimensionality, Measurement Issues, and Determinants." Paper prepared for the European Association of Agricultural Economists (EAAE) Congress, Copenhagen, Denmark, 24–27 August.

Borlaug, Norman (1994). Testimony before the Subcommittee on Foreign Assistance and Hunger, Committee on Agriculture, U.S. House of Representatives, 1 March.

Bradshaw, York W., and Mark J. Schafer (2000). "Urbanization and Development: The Emergence of International Nongovernmental Organizations Amid Declining States," *Sociological Perspectives*, 43 (1, Spring): 97–116.

Bratton, Michael (1989). "The Politics of Government-NGO Relations in Africa," *World Development*, 17: 569–587.

British Broadcasting Corporation (BBC) (2002). "Jose Bove: Profile," BBC Four Documentaries, 20 April, www.bbc.com.uk.

BBC News (2002). "Zambia Refuses GM 'Poison.'" World Edition. 3 September, 13:40 GMT.

British Medical Association (2004). "Genetically Modified Foods and Health: A Second Interim Statement," British Medical Association. London, March.

Brookes, Graham, and Peter Barfoot (2005). "GM Crops: The Global Socioeconomic and Environmental Impact—the First Nine Years 1996–2004," October. Dorchester, UK: PG Economics Ltd.

Buck, Tobias (2007). "Standard Bearer: How the European Union Exports its Laws," *Financial Times*, 10 July, p. 9.

Burnside, Craig, and David Dollar (2000). "Aid, Policies, and Growth," *The American Economic Review*, 90 (4, September): 847–868.

Calvo, Christina M. (1998). *Options for Managing and Financing Rural Transport Infrastructure*. Washington, DC: The World Bank.

Camara, Oumou, and Ed Heinemann (2006). "Overview of the Fertilizer Situation in Africa," background paper prepared for the African Fertilizer Summit, Abuja, Nigeria, 9–13 June.

Carson, Rachel (1962). *Silent Spring*. Boston: Houghton Mifflin.

Center for Food Safety (CFS)(2006). "Market Rejection of Genetically Engineered Foods," Center for Food Safety, August, www.centerforfoodsafety.org.

Centro Internacional de Mejoramiento de Maíz y Trigo (CIMMYT) (2004). "In Quest for Drought-Tolerant Varieties, CIMMYT Sows First Transgenic Wheat Field Trials in Mexico," 12 March, www.cimmyt.org.

Chen, X. (2000). "The Status Quo and Prospect of China's Agricultural Development," remarks at China Development Forum, *China 2010: Charting the Path to the Future.* Beijing, 27–28 March.

Chrispeels, Maarten J. (2004). "Biotechnology and the Poor," *Journal of Plant Physiology,* 124 (September): 3–6.

Clearant (2006). http://www.clearant.com/recombinant.shtml.

Cline, William R. (2007). *Global Warming and Agriculture: Impact Estimates by Country.* Center for Global Development and Peterson Institute for International Economics, Washington, DC.

Cohen, J. I. (2005). "Poor Nations Turn to Publicly Developed GM Crops." *Nature Biotechnology,* 23(1): 27–33.

Connor, Anthony J., Travis R. Glare, and Jan-Peter Nap (2003). "The Release of Genetically Modified Crops into the Environment. Part II. Overview of Ecological Risk Assessment," *The Plant Journal,* 33: 19–46.

Consultative Group on International Agricultural Research (CGIAR) (2007). "Research and Impact: CGIAR and Agricultural Biotechnology," http://www.cgiar.org/impact/agribiotech.html.

CUSO (2001). CUSO Annual Report 2000–2001, Ottawa, http://www.cuso.org/_files/annual_reports/cuso_annual_report_2000_2001.pdf

Daily Telegraph (2002). "Will Their Protests Leave Her Hungry?" *Daily Telegraph* (UK). 23 November, http://www.gene.ch/gentech/2002/Nov/msg00046.html.

Dale, Glenn, and Robert Henry (2003). "Biotechnology for Salt Tolerance and/or Enhanced Water Use in Plants—Interesting Science or a Pathway to the Future?" Centre for Plant Conservation Genetics, Southern Cross University, Australia, www.ndsp.gov.au.

Dawe, D., R. Robertson, and L. Unnevehr (2002). "Golden Rice: What Role Could It Play in Alleviation of Vitamin A Deficiency?" *Food Policy,* 27: 541–560.

Derrick, Jonathan (1977). "The Great West African Drought, 1972–74," *African Affairs,* 76 (305, October): 537–586.

Deutsche Gesellschaft fur Technische Zusammenarbeit (GTZ) (2006). "Capacity Building for an Africa-wide Biosafety System," www.gtz.de/biodiv.

Devarajan, Shantayanan, Andrew Sunil Rajkumar, and Vinaya Swaroop (1999). "What Does Aid to Africa Finance?" Washington, DC: Development Research Group, World Bank.

Devereux, Stephen (1993). "Goats Before Plows: Dilemmas of Household Response Sequencing During Food Shortages," *IDS Bulletin,* 24 (4): 52–59.

Devereux, Stephen, and Trine Naeraa (1996). "Drought and Survival in Rural Namibia," *Journal of Southern African Studies,* 22 (3, September): 421–440.

Dimitri, Carolyn, and Lydia Oberholtzer (2006). "EU and U.S. Organic Markets Face Strong Demand Under Different Policies," *Amber Waves*, Economic Research Service/USDA, 4 (1): 12–19.

Doering, Don S. (2005). "Public-Private Partnership to Develop and Deliver Drought-Tolerant Crops to Food-Insecure Farmers," draft document for discussion. Arlington, VA: Winrock International, 25 April.

Drezner, Daniel W. (2001). "Globalization and Policy Convergence." *International Studies Review*, 3 (1, Spring): 53–78.

Dugger, Celia W. (2007). "World Bank Neglects African Agriculture, Study Says," *The New York Times*, 15 October, p. A3.

——— (2006). "Bush Celebrates Early Victories in Campaign Against Malaria," *The New York Times*, 15 December, p. A7.

Earth Island Institute (EII) (2002). "Summary of Findings: Day of Hunger, Agriculture, Water, and Food Security," World Sustainability Hearing, 30 August, www.earthisland.org/wosh/Day5_Findings.html.

Easterly, William (2007). *The White Man's Burden: Why the West's Efforts to Aid the Rest Have Done So Much Ill and So Little Good*. New York: Penguin.

Economic Research Service (ERS) (2005). "Another Look at Farm Poverty," *Amber Waves*, 3 (4, September): 45.

Egan, Timothy (2006). *The Worst Hard Time: The Untold Story of Those Who Survived the Great American Dust Bowl*. Boston: Houghton Mifflin.

Environmental Negotiations Bulletin (ENB)(1996). "Brief Analysis of the Meeting," *Environmental Negotiations Bulletin*, 9 (48).

Environmental News Service (ENS) (2005). "G8 Leaders Pledge $25 Billion a Year for Africa," Environmental News Service, www.ens-newswire.com.

Etter, Lauren (2007). "DuPont's Biotech Bet," *The Wall Street Journal*, 22 January, p. A10.

EU Food Law Weekly (2006). "Obesity as Big a Killer as Tobacco, Says Madelin," *EU Food Law Weekly*, 268 (1, September), p. 9.

European Council (2005). "EU and Africa: Towards a Strategic Partnership," Ref: CL05–337EN, Statement of the European Council. Brussels, 15–16 December.

European Medicines Agency (EMEA) (2006). http://www.emea.eu.int.

European Union (EU) Research Directorate (2001). "GMOs: Are There any Risks?" Brussels: EU Commission, press briefing, 9 October.

Evenson, Robert E. (1996). "Science for Agriculture: International Perspective," *Asian Journal of Agricultural Economics*, 2: 11–38.

Evenson, Robert E., and Douglas Gollin (2003). *Crop Variety Improvement and Its Effect on Productivity: The Impact of International Research*. Wallingford, UK: CABI Publishing.

Falck-Zapeda, J. B., G. Traxler, and R. Nelson (2000). "Surplus Distribution from the Introduction of a Biotechnology Innovation," *American Journal of Agricultural Economics,* 82 (May): 360–369.

Falcon, Walter P., and Rosamond L. Naylor (2005). "Rethinking Food Security For the 21st Century," *American Journal of Agricultural Economics,* 87(5): 1113–1127.

Falkner, Robert (2006). "The European Union as a 'Green Normative Power'? EU Leadership in International Biotechnology Negotiation," CES Working Paper Series #140, Center for European Studies, Harvard University, Cambridge, MA.

Fan, S., L. Zhang, and X. Zhang (2002). *Growth and Poverty in Rural China: The Role of Public Investment,* International Food Policy Research Institute, Research Report 125. Washington, DC: IFPRI.

Fan, S., P. Hazell, and S. Thorat (2000). "Government Spending, Growth and Poverty in Rural India," *American Journal of Agricultural Economics,* 82(4): 1038–1051.

FAOSTAT (2007). United Nations Food and Agriculture Organization (FAO), Rome. FAOSTAT core production data, www.fao.org/site/340.

FARM-Africa (2007). "Programmes: Smallholder Development & Land Reform," www.farmafrica.org.uk/about.cfm.

Fife-Schaw, C., and G. Rowe (2000). "Extending the Application of the Psychometric Approach for Assessing Public Perceptions of Food Risk; Some Methodological Considerations," *Journal of Risk Research,* 3: 167–179.

Findley, Sally E. (1994). "Does Drought Increase Migration? A Study of Migration from Rural Mali During the 1983–1985 Drought," *International Migration Review,* 28 (3, Autumn): 539–553.

Fitzgerald, Deborah (1993). "Farmers Deskilled: Hybrid Corn and Farmers' Work," *Technology and Culture,* 34 (2, April): 324–343.

Food and Agriculture Organization (FAO) of the United Nations (2006). "People and Populations at Risk," FAO Corporate Document Repository, www.fao.org/docrep/U8480E/U8480E05.htm.

——— (2004). *The State of Food and Agriculture 2003–04: Agricultural Biotechnology: Meeting the Needs of the Poor?* Rome: FAO.

———(2001). *World Markets for Organic Fruit and Vegetables.* International Trade Centre, Technical Centre for Agricultural and Rural Cooperation, Rome: FAO.

Food First (2002). "Food Sovereignty: A Right For All—Political Statement of the NGO/CSO Forum for Food Sovereignty," 14 June. Oakland, CA: Food First.

Food Traceability Report (FTR)(2006). "Wal-Mart Accused of Selling Non-organic Food as 'Organic,'" *Food Traceability Report,* 6 (12, December): 16.

Foodnavigator (2005). "African Food Safety: The Key to Health and Trade?" 4 October, www.foodnavigator.com.

Ford Foundation (2006). "Asset Building for Social Change: Pathways to Large-Scale Impact," New York: Ford Foundation, www.fordfound.org.

French Academy of Medicine (2002). "OGM et sante." Recommendations (Alain Rerat). Communique adopted on 10 December.

French Academy of Sciences (2002). "Genetically Modified Plants." Institut de France, Academie des sciences, Report on Science and technology, 13 (December).

Frewer, L., J. Lassen, B. Kettlitz, J. Scholderer, V. Beekman, and K. G. Berdal (2004). "Societal aspects of genetically modified foods." *Food and Chemical Toxicology*, 42 (2004): 1181–1193.

Frey, K. (1996). *National Plant Breeding Study—I: Human and Financial Resources Devoted to Plant Breeding Research and Development in the United States in 1994* (Special Report No. 98). Ames, IA: Iowa Agricultural and Home Economics Experiment Station.

Friends of the Earth (FoE) (2007). "Environmental Rights Action/Friends of the Earth, Nigeria," "GMO Campaign—FoE Africa," www.eraction.org.

——— (2006). "Illegal Genetically Modified Rice Found," Abuja, Nigeria, 29 November, www.genecampaign.org/Publication/Article/GMtech/GM%20rice-%20Africa%20response.pdf.

——— (2005). "The Resolution Adopted at the FoE-Africa/TWN Conference on GMOs and Africa, Lagos, Nigeria, 21–23 March, www.connectotel.com/gmfood/an160305.txt.

——— (2001). *GMO Contamination Around the World*. Amsterdam, October.

——— (1999). "GMOs: The Case for a Moratorium," http://www.foe.co.uk/resource/reports/gmo_case_for_moratorium.html.

Fromartz, Samuel (2006). *Organic, Inc.* New York: Harcourt.

Fuglie, Keith, Nicole Ballinger, Kelly Day, Cassandra Klotz, Michael Ollinger, John Riley, Utpal Vasavandra, and Jet Yee (1996). *Agricultural Research and Development—Public and Private Investments Under Alternative Markets and Institutions*, Agricultural Economic Report No. AER735, May. Washington, DC: USDA/ERS.

Gaia Foundation (2003). "Meacher Slams US for 'Grotesque Misrepresentation.'" Press release. London: Gaia Foundation, 12 September.

Gakou, Mohamed Lamine (1987). *The Crisis in African Agriculture*. London: Zed Books Ltd.

Galtung, Johan (1971). "A Structural Theory of Imperialism," *Journal of Peace Research*, 8 (2): 81–117.

Gardner, Bruce L. (2002). *American Agriculture in the Twentieth Century: How it Flourished and What it Cost*. Cambridge, MA: Harvard University Press.

Garrett, Laurie (2007). "The Challenge of Global Health." *Foreign Affairs*, 86 (1, January/February): 14–38.

Gaskell, G., N. C. Allum, and S. R. Stares (2003). *Europeans and Biotechnology in 2002: Eurobarometer 58.0.* Brussels: European Commission.

Gaskell, George et al. (2000). "Biotechnology and the European Public." *Nature Biotechnology*, 18: 935–938.

Gaskell, George, Nick Allum, Wolfgang Wagner, Nicole Kronberger, Helge Torgersen, Juergen Hampel, and Julie Bardes (2004). "GM Foods and the Misperception of Risk Perception," *Risk Analysis*, 24 (1): 185–194.

Gaskell, George, Martin Bauer, John Durant, and Nicholas C. Allum (1999). "Worlds Apart? The Reception of Genetically Modified Foods in Europe and the U.S." *Science*, 285: 384–387.

Gee, David, ed. (2002). *The Precautionary Principle in the 20th Century: Late Lessons from Early Warnings.* London: Earthscan.

Genetic Food Alert (2002). "UK Government Minister Condemns 'Wicked' USAID GM Food Policy," Genetic Food Alert UK, 27 November, www.connectotel.com/gmfood/gf271102.txt.

Giddens, Anthony (1999). "Risk and Responsibility," *Modern Law Review*, 62 (1): 1–10.

Gidley, Ruth (2002). "African Crisis Fuels Debate over GM Food," *Reuters Foundation AlertNet*, London, 19 July.

Global Environment Facility (GEF) (2006). "Global Environment Facility Trust Fund: Status of Payments on Contributions as of May 15, 2006," www.gefweb.org.

Global Forum on Agricultural Research (GFAR) (2002). "News from Stakeholders," Issue 5 (December), www.efgar.org/old_newsletters/december2002.

GMO Compass (2006). "Eurobarometer 2006: Europeans Still See More Risks than Benefits," *GMO Compass*, 22 June.

Gopinath, Munisamy, and Terry L. Roe (1997). "Sources of Sectoral Growth in an Economy Wide Context: The Case of U.S. Agriculture." *Journal of Productivity Analysis*, 8: 293–310.

Government of Zambia (2002). "Report of the Factfinding Mission by Zambian Scientists on Genetically Modified Foods," Lusaka, Zambia.

Graham, Edward M. (2000). *Fighting the Wrong Enemy: Antiglobal Activists and Multinational Enterprises.* Washington, DC: International Institute for Economics.

Graig, Augetto (2001). "Consumers Say No to Modified Feeds," *Namibia Economist.*

GRAIN (2006). "Another Silver Bullet for Africa? Bill Gates to Resurrect the Rockefeller Foundation's Decaying Green Revolution." September, www.grain.org/articles/?id-16.

——— (2004). "FAO Declares War on Farmers Not On Hunger," http://www.grain.org/front_files/fao-oopen-letter-june-2004-final-en.pdf.

Greene, Catherine (2006). "U.S. Organic Farm Sector Continues to Expand,"' Economic Research Service/U.S. Department of Agriculture, *Amber Waves,* 4 (2, April): 6.

Grey, Mark A. (2000). "The Industrial Food Stream and its Alternatives in the United States: An Introduction," *Human Organization,* 59 (2, Summer): 143–150.

Grove, Wayne A., and Craig Heinicke (2003). "Better Opportunities or Worse? The Demise of Cotton Harvest Labor, 1949–64," *Journal of Economic History,* 63: 736–767.

GTZ. *See* Deutsche Gesellschaft fur Technische Zusammenarbeit. Haas, Peter (1992). "Introduction: Epistemic Communities and International Policy Coordination," *International Organization,* 46 (1): 1–35.

Hall, Peter, and David Soskice, eds. (2001). *Varieties of Capitalism: The Institutional Foundations of Comparative Advantage.* New York: Oxford University Press.

Hallman, William, Carl Hebden, Helen Aquino, Cara Cuite, and John Lang (2003). *Public Perceptions of Genetically Modified Foods: A National Study of American Knowledge and Opinion.* Publication number RR-1003-004. New Brunswick, NJ: Food Policy Institute, Cook College, Rutgers.

Hand, Eric (2006). "Africa in the Middle of U.S.-EU Biotech Trade War," *St. Louis Post-Dispatch,* 12 December, www.checkbiotech.org.

——— (2005). "Hungry African Nations Balk at Biotech Cassava," *St. Louis Post-Dispatch,* 29 August, www.checkbiotech.org.

HarvestPlus (2006). "Breeding Crops for Better Nutrition." Washington, DC: HarvestPlus, International Food Policy Research Institute, http://www.harvestplus.org/pdfs/brochure.pdf.

——— (2004). "Breeding Crops for Better Nutrition." www.harvestplus.org

Hayami, Yujiro, and Vernon W. Ruttan (1985). *Agricultural Development: An International Perspective.* Baltimore, MD: Johns Hopkins University Press.

Hazell, Peter (2005). "The Role of Agriculture and Small Farms in Economic Development." Prepared for research workshop, "The Future of Small Farms," organized by the International Food Policy Research Institute, Overseas Development Institute, and Imperial College, London. Wye, UK, 26–29 June.

Hazell, Peter, and Lawrence Haddad (2001). "Agricultural Research and Poverty Reduction," Food, Agriculture, and the Environment Discussion Paper 34, International Food Policy Research Institute. Washington, DC: IFPRI.

Helt, Hans Walter (2004). "Are There Hazards for the Consumer When Eating Food from Genetically Modified Plants?" Union of the German Academies of Science and Humanities, Commission on Green Biotechnology. Gottingen: Universitat Gottingen.

Herwig, Lukas (2006). "Tomatoes Against Drought," *Checkbiotech*, 14 November, www.checkbiotech.org.

Heuser, Stephen (2006). "GTC gets surprise boost from EU." *The Boston Globe*, 3 June, p. F7.

Hill, Tony (2004). "Three Generations of UN-Civil Society Relations: A Quick Sketch," Global Policy Forum, www.globalpolicy.org.

Hitt, Greg (2005). "A Kinder, Gentler Wolfowitz at World Bank?" *The Wall Street Journal*, 22 September, p. A4.

Hoban, Thomas J. (2004). "Public Attitudes towards Agricultural Biotechnology," ESA Working Paper 04–09, www.fao.org/es/esa.

Holley, Donald (2000). *The Second Great Emancipation: The Mechanical Cotton Picker, Black Migration, and How They Shaped the Modern South*. Fayetteville: University of Arkansas Press.

Honma, Masayoshi, and Yujiro Hayami (1986). "The Determinants of Agricultural Protection Level: An Econometric Analysis," in Kym Anderson and Yujiro Hayami, *The Political Economy of Agricultural Protection*. Sydney: Allen and Unwin.

Hossain, Ferdaus, Benjamin Onyango, Brian Schilling, William Hallman, and Adesoji Adelaja (2003). "Product Attributes, Consumer Benefits and Public Approval of Genetically Modified Foods," *International Journal of Consumer Studies*, 27 (5, November): 353–365.

Howard, Julie, Valerie Kelly, Mywish Maredia, Julie Stepanek, and Eric W. Crawford (1999). "Progress and Problems in Promoting High External-Input Technologies in Sub-Saharan Africa: The Sasakawa Global 2000 Experience in Ethiopia and Mozambique," selected paper for the Annual Meetings of the American Agricultural Economics Association, Nashville, TN, 8–11 August, http://www.wfc.org.zm.

Huang, J., and S. Rozelle (1996). "Technology Change: Rediscovering the Engine of Growth in China's Rural Economy," *Journal of Development Economics*, 49: 337–369.

Huang, J., R. Hu, C. Fan, C. E. Pray, & S. Rozelle (2002). "*Bt* Cotton Benefits, Costs, and Impacts in China," *AgBioForum*, 5 (4): 153–166. http://www.agbioforum.org.

Huffman, Wallace E., and Robert Evenson (1993). *Science for Agriculture*. Ames: Iowa State University Press.

In Motion Magazine (IMM)(2002). "Interview with Fred Kalibwani of PELUM," *In Motion Magazine*, 6 December, 2002, www.inmotionmagazine.com/global/fk1.html.

Intergovernmental Panel on Climate Change (IPCC) (2007). *Climate Change 2007: Climate Change Impacts, Adaptation, and Vulnerability*. Working Group II

Contribution to the IPCC Fourth Assessment Report: Climate Change 2007. Geneva, April 6.

International Council for Science (2003). *New Genetics, Food and Agriculture: Scientific Discoveries—Societal Dilemmas.* www.icsu.org.

International Federation of Organic Agriculture Movements (IFOAM) (2006). "FAO Shows Interest in Exploring the Potential of Organic Agriculture for Food Security," News/Press, IFOAM, 2 November. www.ifoam.org.

International Food Information Council (IFIC) (2005). "Food Biotechnology: Survey Questionnaire 6/1/05," IFIC Background, June.

International Food Policy Research Institute (IFPRI) (2006). "How Will Agriculture Adapt to a Shifting Climate?" *IFPRI Forum,* International Food Policy Research Institute, Washington, DC. Impact assessment of *Bt* corn in the Philippines, 2004. Terminal Report.

——— (2004). "Funding Africa's Farmers," *IFPRI Forum,* International Food Policy Research Institute, March. Washington, DC: IFPRI.

International Institute for Environment and Development (IIED) (2006). "A Citizens Space for Democratic Deliberation on GMOs and the Future of Farming in Mali," 24 January. http://www.iied.org/NR/agbioliv/ag_liv_projects/GMOCitizenJury.html.

International Institute for Sustainable Development (IISD) (2007). "Biosafety Policy Brief: African Regional Coverage Project," 5, (1, 7 February).

International Service for the Acquisition of Agri-Biotech Applications (ISAAA) (2002). "*Bt* Cotton in South Africa," http://www.isaaa.org/kc.

Investors Business Daily (2006). "Let Them Eat Cake." www.investors.com/terms/reprints.asp.

IRIN News (2003). "Zambia: WFP Delivers Non-GM Food Aid," IRIN News, 30 January, http://www.irinnews.org/report.asp?ReportID=32014.

Jaffe, Gregory (2006). "Regulatory Slowdown on GM Crop Decisions," *Nature Biotechnology,* 24 (7): 748.

——— (2005). "Withering on the Vine?: Will Agricultural Biotechnology's Promise Bear Fruit?" Center for Science in the Public Interest, 2 February. www.cspinet.org.

James, Clive (2006). *Global Status of Commercialized Biotech/GM Crops: 2006.* International Service for the Acquisition of Agri-Biotech Applications (ISAAA), Brief No. 35. ISAAA: Ithaca, NY.

Jasanoff, Sheila (2005). *Designs on Nature: Science and Democracy in Europe and the United States.* Princeton: Princeton University Press.

Jayne, T. S. (1994). "Do High Food Marketing Costs Constrain Cash Crop Production? Evidence from Zimbabwe," *Economic Development and Cultural Change,* 42 (2): 387–402.

Jayne, T. S., D. Mather, and E. Mghenyi (2005). "Smallholder Farming in Difficult Circumstances: Policy Issues in Africa," paper presented at "The Future of Small Farms," International Food Policy Research Institute and Overseas Development Institute, and Imperial College, London. Wye, UK, 26–29 June.

Jesuit Centre for Theological Reflection (JCTR) (2002). "What Is the Impact of GMOs on Sustainable Agriculture in Zambia?" Jesuit Centre for Theological Reflection, Lusaka, Zambia, 15 August.

Johnson, Andrea (2006). "Agri-Tech: Drought Tolerance Is the Next Step in Developing Biotech Crop Traits," *Lee Agri-Media,* 25 October. www.tristateneighbor.com/articles/2006.

Jones, Charles I. (2002). "Sources of U.S. Economic Growth in a World of Ideas," *American Economic Review,* 92 (1, March): 220–239.

Josling, Tim, and Stefan Tangermann (2006). "Distortions to Agricultural Incentives in Western Europe," draft of Agricultural Distortions Research Project Working Paper, October. Washington, DC: World Bank.

Juma, Calestous, and Ismail Serageldin (2007). *Freedom to Innovate: Biotechnology in Africa's Development.* African Union (AU) and New Partnership for Africa's Development (NEPAD), April. www.africa-union.org and www.nepadst.org.

Kaufmann, D., A. Kraay, and M. Mastruzzi (2007). *Governance Matters VI: Governance Indicators for 1996–2006.* Policy Research Paper No. 4280, Washington, DC: World Bank.

Keck, Margaret E., and Kathryn Sikkink (1998). *Activists Beyond Borders: Advocacy Networks in International Politics.* Ithaca, NY: Cornell University Press.

Kessler, Charles, and Ioannis Economidis, eds. (2001). *EC-Sponsored Research on Safety of Genetically Modified Organisms: A Review of Results.* Luxembourg: Office for Official Publications of the European Communities.

Kigotho, Wachira (2005). "Decades of Drought Predicted for Southern Africa," June 2. www.scidev.net/news.

Kilman, Scott, and Roger Thurow (2002). "African Famine," *Wall Street Journal,* 3 December, p. 1.

Kindleberger, Charles P. (1951). "Group Behavior and International Trade," *Journal of Political Economy,* 50 (February): 30–46.

Kirsten, J., and M. Gouse (2003). "The Adoption of and Impact of Agricultural Biotechnology in South Africa," in N. Kalaitzandonkes, ed., *The Economic and Environmental Impact of Agbiotech: A Global Perspective.* New York: Kluwer Academic Press, Plenum Publishers.

Kjaernes, Unni, Arne Dulsrud, and Christian Poppe (2006). "Contestation over Food Safety: The Significance of Consumer Trust," pp. 61–95 in Christopher Ansell and David Vogel, eds., *What's the Beef? The Contested Governance of European Food Safety.* Cambridge, MA: MIT Press.

Klotz-Ingram, C., and K. Day-Rubenstein (1999). "The Changing Agricultural Research Environment: What Does it Mean for Public-Private Innovation?" *AgBioForum,* 2 (1): 24–32.

Kogan, Lawrence A. (2005). *Exporting Precaution: How Europe's Risk-Free Regulatory Agenda Threatens American Free Enterprise.* Washington, DC: Washington Legal Foundation.

Krishnakumar, Asha (2004). *Frontline,* 21 (12, June): 5–18.

Lappe, Frances Moore (1971). *Diet for a Small Planet.* New York: Ballantine Books.

Lappe, Frances Moore, and Joseph Collins (1977). *Food First: Beyond the Myth of Scarcity.* Boston: Houghton Mifflin.

Lear, Linda (2002). "Introduction" to Rachel Carson, *Silent Spring,* pp. x–xix, Fortieth Anniversary Edition, Boston: Houghton Mifflin.

Lesseps, Roland (2003). "Genetic Engineering Evaluated from the Perspective of Christian and Ignatian Creation Spirituality," Rome, www.sjweb.info/sjs.

Lewanika, Mwananyanda (2002). "Norway-Zambia Collaboration in Biosafety Capacity Building," www.biodiv.org/doc/meetings/bs/bscmcb-02-inf-02-en.pdf.

Lichfield, John (2007). "Police Tear-Gas Farmers in Clash over French GM Crops," *Belfast Telegraph,* 27 August. www.belfasttelegraph.co.uk/news/world-news/article2898542.ece.

Lipton, Michael (2005). *The Family Farm in a Globalizing World.* 2020 Discussion Paper 40, June. Washington, DC: IFPRI.

——— (1977). *Why Poor People Stay Poor.* Cambridge, MA: Harvard University Press.

Lipton, Michael, and Robert Paarlberg (1990). "The Role of the World Bank in Agricultural Development in the 1990s," International Food Policy Institute (IFPRI). Washington, DC.

Little, Peter D., M. Priscilla Stone, Tewodaj Mogues, A. Peter Castro, and Workneh Negatu (2004). "Moving in Place: Drought and Poverty Dynamics in South Wollo, Ethiopia," 8 December. Madison: University of Wisconsin. www.basis.wisc.edu/live/persistent%20poverty.

Lovelock, James (2006). *The Revenge of Gaia.* New York: Basic Books.

Maathai, Wangari (2004). "Nobel Lecture," 10 December. Oslo, Norway. www.nobelprize.org.

Maine Organic Farmers and Gardeners Association (MOFGA)(2006). "What Does MOFGA Mean? Supporting Maine's Family Farmers since 1971," *Edible Coastal Maine* (Summer): 21.

Mantell, Katie (2002). "WHO Urges Africa to Accept GM Food Aid," Science and Development Network, 30 August. www.scidev.net/News.

Marris, Claire (2001). "Public Views on GMOs: Deconstructing the Myths," *EMBO Reports*, 2 (7): 545–548.

Martin, Paul (2002). "Greenpeace, Zambia Reject U.S. Claim." *Washington Times*, 31 August. www.connectotel.com/gmfood/wt310802.txt.

McCalla, Alex F. (1974). "Review of Hightower, Jim, *Hard Tomatoes, Hard Times: A Report of the Agribusiness Accountability Project on the Failure of America's Land Grant College Complex,*" *American Journal of Agricultural Economics*, 56 (2, May): 461–462.

McCann, James C. (1999). "Climate and Causation in African History," *The International Journal of African Historical Studies*, 32, (2/3): 261–279.

Mead, P. S., L. Slutsker, V. Dietz, et al. (1999). "Food Related Illness and Death in the United States." *Emerging Infectious Disease*, 5: 607–625.

Meinzen-Dick, Ruth, Michelle Adato, Lawrence Haddad, and Peter Hazell (2004). "Science and Poverty: An Interdisciplinary Assessment of the Impact of Agricultural Research," International Food Policy Research Institute. Washington, DC: IFPRI.

Melcer, Rachel (2004). "Scientists Zero-in on Drought Resistant Crops," *St. Louis Post-Dispatch*, 1 June, www.agbioworld.org.

Mellon, Margaret (2001). "Does the World Need GM Foods?" GM Food Safety Q&A, *Scientific American* (April): 64–65.

Miller, Henry I., and Gregory Conko (2004). *The Frankenfood Myth: How Protest and Politics Threaten the Biotech Revolution*. Westport, CT: Praeger.

Mitchell, Lorraine (2004). "U.S. and EU Consumption Comparisons," pp. 49–65 in *U.S.-EU Food and Agriculture Comparisons/WRS-04-04*. Economic Research Service, USDA.

Mitchner, Brandon (2003). "Rules, Regulations of Global Economy Increasingly Set in Brussels." *Wall Street Journal*, 23 April.

Monsanto/AATF (2007). "Combining Breeding and Biotechnology to Develop Drought Tolerant Maize for Africa: A Proposal to the Bill and Melinda Gates Foundation," submitted by Monsanto Company and the African Agricultural Technology Foundation, 25 May 25. St. Louis.

Monsanto Company (2006). "Reflections of a Science Pioneer: Rob Horsch Says Goodbye to Monsanto," www.monsanto.com/monsanto/layout/reflections/rob_horsch.asp.

——— (2005). "Corn Yield—Drought Tolerance," www.monsanto.com/monsanto/content/investor/financial/presentations/2005/poster_01-18-05a.pdf.

——— (2003). "Progress in the Product Pipeline," *Annual Report 2003*, pp. 12–

13, www.monsanto.com/monsanto/content/media/pubs/2003/2003_Annual_Report_Pipeline.pdf.
——— (1999). "Gene Protection Technologies: A Monsanto Background Statement," pp. 127–128 in J. Janick, ed., *Perspectives on New Crops and New Uses.* Alexandria, VA: ASHS Press.
Moon, Wanki, and Siva K. Balasubramanian (2001). "Public Perceptions and Willingness-to-pay a Premium for Non-GM Foods in the US and UK," *AgBioForum,* 4(3&4): 221–231.
Mugabe, J., et al. (2000). *Global Biotechnology Risk Management: A Profile of Policies, Practices, and Institutions.* Nairobi: UNEP and ACTS.
Munoz, Sara S. (2007). "For Sale: Condo W/Chicken Coop," *Wall Street Journal,* May 17, p. D1.
Murphy, Sophia (2006). "Is a Green Revolution for Africa the Answer?" IATP Commentary, 27 October, Minneapolis: IATP.
Mwenda, Andrew M. (2005). "Foreign Aid Sabotages Reform," *International Herald Tribune,* 8 March.
National Academy of Sciences (NAS)(2004). *Safety of Genetically Engineered Foods: Approaches to Assessing Unintended Health Effects.* Washington, DC: NAS.
Natsios, Andrew (2006a). "Focus," *Foreign Service Journal* (June): 22–24.
——— (2006b). "Hunger, Famine, and the Promise of Biotechnology," in Jon Entine, ed., *Let Them Eat Precaution.* Washington, DC: AEI.
Network for Ecofarming in Afria (NECOFA) (2005). "Brief Report on NECOFA Kenya Activities 2004/05," *NECOFA Newsletter,* 7 (4). www.necofa.org.
New Agriculturalist (2005). "The Quest for Drought Tolerance," 10 March. www.checkbiotech.org.
NewScientist (2003). "Zambia's GM Food Fear Traced to UK," 29 January. www.newscientist.com/article.ns?id=dn3317.
Ngandwe, Talent (2005). "Zambia Builds High-Tech Lab to Detect GM Food Imports," *SciDev.Net,* 13 May. www.scidev.net.home.
NGO Forum (1996). "Profit for Few or Food For All." World Food Summit, Rome, 17 November.
NGO/CSO Forum (2002). "'Profit For Few or Food For All' Revisited Five Years Later." 2002 Rome NGO/CSO Forum for Food Sovereignty, Rome, 10 June.
Nicholson, Sharon (2000). "The Nature of Rainfall Variability Over Africa on Time Scales of Decades to Millennia," *Global and Planetary Change,* 26 (2000): 137–158.
Nicholson, Simon (2006). "Risk as Rhetoric: Constructing Zambia's GM Foods Policy," International Studies Association, San Diego, CA, 22–26 March.

Nordlee, J. S., S. Taylor, J. Townsend, L. Thomas, and R. Bush (1996). "Identification of a Brazil-nut Allergen in Transgenic Soybeans." *New England Journal of Medicine*, 334: 688–692.

Nuffield Council on Bioethics (1999). *Genetically Modified Crops: The Ethical and Social Issues*. London, May.

Oestreich, Dean (2004). "Business Review Series," Pioneer-DuPont Company, September 2004. http://media.corporate-ir.net/media_files/NYS/DD/presentations/Oestreich1.pdf.

Office of Technology Assessment (OTA) (1992). *A New Technological Era for American Agriculture*. Washington, DC: GPO.

Offutt, Susan, and Craig Gundersen (2005). "Farm Poverty Lowest in U.S. History," *Amber Waves*, 3 (4), Economic Research Service, USDA.

Omamo, Steven, et al. (2006). *Strategic Priorities for Agricultural Development in Eastern and Central Africa*. Research Report 150, International Food Policy Research Institute, Washington, DC: IFPRI.

Orden, David, Robert Paarlberg, and Terry Roe (1999). *Policy Reform in American Agriculture: Analysis and Prognosis*. Chicago: The University of Chicago Press.

Organisation for Economic Co-operation and Development (OECD) (2000). "GM Food Safety: Facts, Uncertainties, and Assessment, Rapporteurs' Summary." The OECD Edinburgh Conference on the Scientific and Health Aspects of Genetically Modified Foods, 28 February–1 March.

Oxfam International (2006). "Causing Hunger: An Overview of the Food Crisis in Africa." Briefing Paper 91, July. Oxford, UK: Oxfam.

Oxford Policy Management (2007). "Uganda: Agriculture Sector Public Expenditure Review." Oxford, UK. www.opml.co.uk.

Paddock, William, and Paul Paddock (1967). *Famine, 1975! America's Decision: Who Will Survive?* Boston: Little Brown.

Pardey, Philip G., Julian M. Alston, and Roley R. Piggott (2006). *Agricultural R&D in the Developing World: Too Little, Too Late?* Washington, DC: IFPRI.

Pardey, Philip G., Nienke Beintema, Steven Dehmer, and Stanley Wood (2006). *Agricultural Research: A Growing Global Divide?* Agricultural Science, the Technology Indicators Initiative, International Food Policy Research Institute, August.Washington, DC: IFPRI.

Park, Sungun, Jisheng Li, Jon K. Pittman, Gerald A. Berkowitz, Haibing Yang, Soledad Undurraga, Jay Morris, Kendal D. Hirschi, and Roberto A. Gaxiola (2005). "Up-regulation of a H+-pyrophosphatase (H+-PPase) as a Strategy to Engineer Drought-resistant Crop Plants," *Proceedings of the National Academy of Sciences*, 102 (52): 18830–18835.

Parr, Doug (2002). "Foreword: Food Security for All the World's People," pp. 4–9 in Nicholas Parrott and Terry Marsden, *The Real Green Revolution: Organic and Agroecological Farming in the South*. London: Greenpeace Environmental Trust.

Parrott, Nicholas, and Terry Marsden (2002). *The Real Green Revolution: Organic and Agroecological Farming in the South.* London: Greenpeace Environmental Trust.

Patterson, L. A. (2000). "Biotechnology Policy," in H. Wallace and W. Wallace, eds., *Policy-Making in the European Union.* Oxford: Oxford University Press.

Pei, Z-M., M. Ghassemian, C. M. Kwak, P. McCourt, and J. L. Shroder (1998). "Role of Farnesyltransferase in ABA Regulation of Guard Cell Anion Channels and Plant Water Loss," *Science,* 282: 287–290.

Pellegrineschi, Alessandro (2004). "Drought Tolerant Crops and Transgenic Breeding: Just a Utopian Vision?" *PBI Bulletin,* issue 2.

PELUM (2004). "Biosafety Bill Endangers Kenya," media release, Nairobi, 2 September.

Performance Plants (2005). "Performance Plants Solves Decades-Old Drought Tolerance Puzzle," press release, 19 July. www.performanceplants.com.

Peterson, Brad (2006). "Our Engineered Food Supply," *Seed Magazine,* 3 March. http://www.seedmagazine.com.

Pew Initiative on Food and Biotechnology (2005). "Public Sentiment about Genetically Modified Food, November Update." http://pewagbiotech.org/research/2005update/2.php.

Phiri, Brighton (2002). "US Comes Under Attack Over GMOs," *The Post,* Zambia, 13 August. http://connectotel.com/gmfood/pz130802.txt.

Pingali, Prabhu, Kostas Stamoulis, and Randy Stringer (2006). "Eradicating Poverty and Hunger: Towards a Coherent Policy Agenda," Working Paper No. 06–01, United Nations Food and Agriculture Organization. Rome: FAO.

Pinstrup-Andersen, Per, Rajul Pandya-Lorch, and Suresh Babu (1997). "2020 Vision for Food, Agriculture and the Environment in Southern Africa," in Lawrence Haddad, ed., *Achieving Food Security in Southern Africa.* Washington, DC: International Food Policy Research Institute (IFPRI).

Pioneer (2007). "Drought Tolerance—Corn." http://www.pioneer.com/CMRoot/Pioneer/research/pipeline/spec_sheets/Drought.pdf.

Plotkin, Robert (2003). "A Zambian Pseudo-Famine," *Point Reyes Light,* Archive, 10 May. www.ptreyeslight.com.

Pollan, Michael (2006). *The Omnivore's Dilemma: A Natural History of Four Meals.* New York: The Penguin Press.

Postma, William (1994). "NGO Partnership and Institutional Development: Making It Real, Making It Intentional." *Canadian Journal of African Studies,* 28 (3): 447–471.

Potrykus, Ingo (2007). "Comment of Ingo Potrykus on World Development Report 2008: Agriculture for Development," 22 April. www.agbioworld.org.

Pratt, Sean (2005). "GM Canola Handles Drought," *Western Producer,* 28 October. www.agbios.com.

Pray, C. E., P. Bengali, and B. Ramaswami (2004). "Costs and Benefits of Biosafety Regulation in India: A Preliminary Assessment." Paper presented at the 8th ICABR Conference, 8–11 July, Ravello, Italy.

Priest, Susanna Hornig, Heinz Bonfadelli, and Maria Rusanen (2003). "The 'Trust Gap' Hypothesis: Predicting Support for Biotechnology Across National Cultures as a Function of Trust in Actors," *Risk Analysis*, 23 (4): 751–766.

Putnam, Judy, Jane Allshouse, and Linda Scott Kantor (2002). "U.S. Per Capita Food Supply Trends: More Calories, Refined Carbohydrates, and Fats," Economic Research Service, USDA, *FoodReview*, 25 (3): 2–15.

Rasmussen, Wayne D., and Paul Steven Stone (1982). "Toward a Third Agricultural Revolution," *Proceedings of the Academy of Political Science*, 34 (3): 174–185.

Religion & Ethics Newsweekly (REN) (2004). "Zambia: Genetically Modified Foods," cover story, *Religion & Ethics Newsweekly*, 23 January. www.pbs.org/wnet/religionandethics/index_flash.html.

Rice, Andrew (2006). "Misery Chic," *New York Times Magazine*, 10 December, p. 59.

Robinson, Guy (2003). "Communique to Zambia Agricultural Investment Promotion Conference," Zambia National Farmers Union, November. Lusaka, Zambia.

Rodin, Judith (2006). World Food Day address to United Nations Food and Agriculture Organization, 18 October. Rome: FAO.

Rose, Richard (1993). *Lesson-Drawing in Public Policy: A Guide to Learning Across Time and Space*. Chatham, NJ: Chatham House.

Rosegrant, Mark W., Sarah A. Cline, Weibo Li, Timothy B. Sulser, and Rowena A. Valmonte-Santos (2005). *Looking Ahead: Long-Term Prospects for Africa's Agricultural Development and Food Security*. 2020 Discussion Paper 41, Washington, DC: International Food Policy Research Institute.

Rosset, Peter (2006). "Bill Gates/Rockefeller Programs to Alleviate Hunger in Africa Via GMOs and Industrialized Farming Are Doomed," *Seattle Post-Intelligencer*, 22 September.

——— (2000). "Lessons from the Green Revolution," Oakland, CA: Food First/Institute for Food and Development Policy.

Royal Society (2003). "Where Is the Evidence that GM Foods Are Inherently Unsafe, Asks Royal Society," press release, Royal Society, 8 May.

Ruttan, Vernon W. (2004). "Controversy About Agricultural Technology Lessons from the Green Revolution," *International Journal of Biotechnology*, 6 (1): 43–54.

——— (2001). *Technology, Growth, and Development: An Induced Innovation Perspective*. New York: Oxford University Press.

Ruttan, Vernon W., Hans P. Binswanger, Yujiro Hayami, William W. Wade, and Adolf Weber (1978). "Factor Productivity and Growth: A Historical Interpretation," in Hans P. Binswanger and Vernon W. Ruttan, eds., *Induced Innovation: Technology, Institutions and Development.* Baltimore: Johns Hopkins University Press, chap. 3.

Sachs, Jeffrey D., John McArthur, Guido Schmidt-Traub, Margaret Kruk, Chandrika Bahadur, Michael Faye, and Gordon McCord (2004). "Ending Africa's Poverty Trap," *Brookings Papers on Economic Activity,* 1: 117–240.

Sanvido, Olivier, Jorg Romeis, and Franz Bigler (2007). *Advanced Biochemical Engineering/Biotechnology,* 107 (31 March): 235–278. Published online.

Sawahel, Wagdy (2004). "Egyptian Scientists Produce Drought-Tolerant GM Wheat," *SciDevNet,* 14 October.

Schmitz, Andrew, and David Seckler (1970). "Mechanized Agriculture and Social Welfare: The Case of the Tomato Harvester," *American Journal of Agricultural Economics,* 52 (4, November): 569–577.

Schultz, T. W. (1964). *Transforming Traditional Agriculture.* New Haven, CT: Yale University Press.

Scott, Christina Scott (2004). "Angola Rejects GM Food Aid," *SciDevNet,* 2 April. www.scidev.net/News/index.cfm?fuseaction=readnews&itemid=1007&language=1.

SeedQuest (2005). "DuPont Donates Technology Valued at US$4.8 million to Africa Nutritionally Enhanced Sorghum Project," SeedQuest News Section, 1 July, Des Moines.

Selin, Henrik, and Stacy D. VanDeveer (2006). "Raising Global Standards: Hazardous Substances and E-Waste Management in the European Union," *Environment,* 48 (10): 6–17. http://www.heldref.org/env.php.

Shapin, Steven (2006). "Paradise Sold," *New Yorker,* 15 May, pp. 84–88.

Shoemaker, Robbin, D. Demcey Johnson, and Elise Golan (2003). "Consumers and the Future of Biotech Foods in the United States," USDA, Economic Research Service, *Amber Waves,* 1 (5, November): 30–36.

Silver, Lee (2006). *Challenging Nature.* New York: Harper Collins.

Smaling, Eric, Moctar Toure, Nico de Ridder, Nteranya Sanginga, and Henk Breman (2006). "Fertilizer Use and the Environment in Africa: Friends or Foes?" Background paper, African Fertilizer Summit, Abjua, Nigeria, 9–13 June.

Smil, Vaclav (1997). "Global Population and the Nitrogen Cycle," *Scientific American Magazine,* July, pp. 76–81.

Smith, Lisa C., and Lawrence Haddad (2000). *Overcoming Child Malnutrition in Developing Countries,* Discussion Paper 30 (February). Washington, DC: International Food Policy Research Institute (IFPRI).

Spielman, D. J., J. Cohen, and P. Zambrano (2006). "Will Agbiotech Applications Reach Marginalized Farmers? Evidence from Developing Countries," *AgBioForum,* 9 (1): 23–30. www.agbioforum.org.

Swaminathan, M. S. (1994). *CIMMYT in 1992: Poverty, the Environment, and Population Growth, the Way Forward.* Mexico City: CIMMYT.

Tewolde, B. G. Egziabher, and Susan Burnell Edwards (2005). "Can Organic Farming Feed the World?" Lecture given in July 2005 on behalf of the Soil Association. www.soilassociation.org.

Third World Network (2002). "Don't Pressure Hungry Peoples to Accept GM Food Aid, Open Letter to the Government of the United States of America, the World Food Programme, the World Health Organization, and the Food and Agriculture Organization," Johannesburg, 2 September. www.twnside.org.sg/title/geletter.htm.

Thirtle, C., L. Lin, and J. Piesse (2003). "The Impact of Research-Led Agricultural Productivity Growth in Africa, Asia and Latin America," *World Development,* 31: 1959–1975.

Tokarick, Stephen (2003). "Measuring the Impact of Distortions in Agricultural Trade in Partial and General Equilibrium," Working Paper WP/03/110. Washington, DC: International Monetary Fund.

Torgersen, Helge (2000). "Lessons of the Past of Biotech in Europe," Institute of Technology Assessment, Austrian Academy of Sciences, Vienna. www.oeaw.ac.at/ita/ebene5/HTPruhoshort.pdf.

Tracy, Michael (1989). *Government and Agriculture in Western Europe 1880–1988,* 3d ed. New York: New York University Press.

Trewavas, A. J. (2004). "A Critical Assessment of Organic Farming and Food Assertions with Particular Respect to the UK and the Potential Environmental Benefits of No-Till Agriculture," *Crop Protection,* 23: 757–781.

Tsiko, Sifelani (2006). "Southern Africa: Boost Capacity for GMO Testing SADC States Told," 8 September. www.checkbiotech.org.

Tuberosa, Roberto, and Silvio Salvi (2006). "Genomics-based Approaches to Improve Drought Tolerance of Crops," *Trends in Plant Science,* 11 (8): 405–412.

Tuskegee University (2001). "Lessons Learned on Rural Development and Economic Performance in Africa," Tuskegee University Workshop, 19–21 April, Atlanta, Georgia. Tuskegee, AL: Tuskegee University.

United Nations Development Programme (UNDP)(2004). *UNDP Annual Report,* www.undp.org/annualreports/2004.

United Nations Environment Programme (UNEP) (2006). "A Comparative Analysis of Experiences and Lessons from UNEP-GEF Biosafety Projects," prepared by the UNEP-GEF Biosafety Unit, as of December 2006. GEF Secretariat. Washington, DC.

——— (2005). "Phase 3 Toolkit Module Part (i): Developing the Regulatory Regime." www.unep.org.

United Nations Food and Agriculture Organization (2006). *The State of Food Insecurity in the World.* Food and Agriculture Organization of the United Nations. Rome: FAO.

United States Agency for International Development (USAID)(2003). *Strategic Plan for Fiscal Years 2004–2009: Security, Democracy, Prosperity.* Washington, DC: U.S. Department of State and U.S. Agency for International Development.

United States Department of Agriculture (USDA) (2006a). "Adoption of Genetically Engineered Crops in the United States," Economic Research Service, 18 August. http://www.ers.usda.gov/Data/BiotechCrops/.

——— (2006b). Global Agriculture Information Network (GAIN). GAIN Report Number EG6020, 20 July. www.fas.usda.gov/gainfiles/200607/146208389.doc.

——— (2004a). *Food Security Assessment.* Economic Research Service. Agriculture and Trade Report No. GFA15, 20 May.

——— (2004b). "How Many U.S. Households Face Hunger . . . and How Often?" *Amber Waves,* Economic Research Service/USDA, 2 (1, February): 7.

Van de Giesen, N. C., T. Berger, W. Laube, C. Rodgers, and Paul L. G. Vlek (2005). "Hydrological Potential, Economic Evaluation, and Institutional Constraints: Decision Support for Irrigation Development in the Volta Basin." *Geophysical Research Abstracts,* 7, 05179, European Geosciences Union.

Van Zanden, J. L. (1991). "The First Green Revolution: The Growth of Production and Productivity in European Agriculture, 1870–1914," *The Economic History Review,* New Series, 44 (2): 215–239.

Vogel, David (1995). *Trading Up: Consumer and Environmental Regulation in a Global Economy.* Cambridge, MA: Harvard University Press.

Wamboga-Mugirya, Peter (2006). "Uganda: Country Gets GM Testing Tools," *SciDev.net,* 15 December. www.checkbiotech.org.

Weise, Elizabeth (2005). "Drought Resistant Corn Sprouts," *USA Today,* 28 July. www.checkbiotech.org.

Wenzel, Wayne (2006). "Advances in Pinpointing Plant Genes for Drought-Tolerance," 4 September. www.checkbiotech.org.

Western Organization of Resources Councils (WORC) (2006). "Potential Market Impacts from Commercializing Roundup Ready Wheat," September. www.worc.org/pdfs/Market%20Risks%20Update%20Final208–06.pdf.

Williams, Robert G. (1986). *Export Agriculture and the Crisis in Central America.* Chapel Hill: The University of North Carolina Press.

Williamson, Claire (2007). "Is Organic Food Better for Our Health?" *Nutrition Bulletin,* 32 (2): 104–108.

Winter, Carl K., and Sarah F. Davis (2006). "Organic Foods." *Journal of Food Science*, 7 (9): 117–124.

Wolson, Rosemary (2006). "Appendix F: South Africa Case Study," *Introducing Biofortified Foods into Developing Countries: An Analysis of the Political Landscape.* A Report to the Bill and Melinda Gates Foundation, Belfer Center for Science and International Affairs, Harvard University, Cambridge, MA.

Women for Change (WFC)(2007). "Women for Change: WFC Networking." www.wfc.org/zm/networks.htm.

World Bank (2008). *World Development Report 2008: Agriculture for Development.* Washington, DC: World Bank. www.worldbank.org.

——— (2006a). *World Development Indicators 2006* www.worldbank.org.

——— (2006b). "World Bank Lending to Borrowers in Africa by Theme and Sector." *World Bank Annual Report 2006.* Washington, DC: World Bank. www.worldbank.org.

——— (2005). "The International Assessment of Agricultural Science & Technology for Development (IAASTD): At a Glance," March, www.agassessment.org.

——— (2003). *World Development Indicators 2003.* Washington, DC: World Bank.

——— (2001). *World Development Report 2000/2001: Attacking Poverty.* New York: Oxford University Press.

——— (2000). "Poverty and Climate Change," in *Environment Matters: Annual Review July 1999–June 2000.* Washington, DC: World Bank, pp. 22–25.

——— (1999). "Agricultural Biotechnology and Rural Development: Issues and Options for World Bank Support to Research and Capacity Building." Draft discussion paper, 26 May. Washington, DC: World Bank.

World Resources Institute (WRI) (2006). "Earth Trends > Agriculture and Food > Variable." http://earthtrends.wri.org/text/agriculture-food/variable-177.html.

——— (1992). *World Resources 1992–93.* New York: Oxford University Press.

World Summit on Sustainable Development (WSSD) (2002). "Johannesburg Recommendations on Sustainable Development Adopted at the 17th Plenary Meeting of the World Summit on Sustainable Development," Johannesburg, 4 September.

World Trade Organization (WTO) (2006). "Dispute Settlement: Reports Out on Biotech Disputes," 29 September. www.wto.org/english/news_e/news06_e/291r_e.htm.

Yale Center for Environmental Law and Policy (2006). *2006 Environmental Performance Index.* www.yale.edu/epi.

Yeats, Alexander J., Azita Amjadi, and Ulrich Reincke (1996). "Did External Barriers Cause the Marginalization of Sub-Saharan Africa in World Trade?" Washington, DC: World Bank.

Yorobe, J. M. Jr., and C. B. Quicoy (2004). "Economic Impact of *Bt* corn in the Philippines." *The Philippine Agricultural Scientist,* 89 (3): 258–267.

Zencey, Eric (2002). Review of "The Art of the Commonplace: the Agrarian Essays of Wendell Berry," ed. and intro. by Norman Wirzba. *Nation,* 275 (1, July).

INDEX